The Theory and Practice of

Cylinder Head Modification

by David Vizard

First published as hardback in March 1971
This paperback edition February 1973

ISBN O-85113-066-6

Edited and produced by Dave Hipperson, Dave Jevons and Ian Ward.

Published by
Interauto Book Company Limited
Bercourt House
51 York Road, Brentford, Middx., England

Printed in England by
HGA Printing Company Ltd.
Brentford, Middlesex

Contents

Foreword

Formal text books on Internal Combustion Engines are usually theoretical appreciations of the thermal and fluidic systems involved in heat engines. The present book is more concerned with the practical significance of these subjects as applied to extracting higher specific power outputs from a particular engine. There is an informal treatment of the basic theory, apart from the description of the limitations to maximum possible cycle efficiency. The book provides an introduction to useful modern development methods not previously published in a single reference text.

The book was written primarily for the novice engine tuner but it contains such a wealth of information, particularly with respect to cylinder head gas-flowing, that it will also be invaluable to the professional. The theoretical knowledge required of the reader is a minimum consistent with an intelligent appreciation of the subject.

The examples, being intended for the practical engine tuner are comprehensive and include suggestions for development beyond those described within the text. Indeed they suggest to the inquiring mind avenues for further research in the field.

T. R. Boyce, B.A.Sc., M.A.Sc., Ph.D., D.I.C.

Introduction

The art of modifying production cylinder heads to give increased power output is one which many people lay claim to but which few, in fact, possess. Terms such as 'gas flowed', 'high compression' etc are often heard but ninety-nine times out of a hundred the particular head in its modified form has never been near a gas flow machine or even had a mathematical analysis of the gas flow done on it. As for a raised compression ratio, this is thought by a great many to be the total extent or nearly so, of head modifying, a point which is, in effect, quite far from the truth. There is a great deal more to modifying a head than just raising the CR or even polishing the head and then raising the CR.

Other points of error which are often encountered are the fitting of larger valves when they are either (a) not needed or (b) other aspects of head modifications have not been considered. The net result of such a move is to achieve less power increase than could have otherwise been gained, or even to reduce the power below that of a standard output.

Ports which are too large can have a measurable adverse effect on power and there are numerous production heads which have too large ports as standard. Further enlargement above and beyond this serves only to make the situation worse.

Taking an all-round view of things, there are many pitfalls for the unwary head modifier to fall into. To acquire all the facts necessary to avoid such pitfalls will mean many a long hour spent in study. In fact so much time can be spent acquiring the necessary knowledge that it is hardly worth it from the home tuner's point of view, since he may only modify one or two heads a year. Bearing this in mind, this book has been laid out to give a practical view of head modifying rather than to present a whole load of proven theory from which the reader may draw his own conclusions. The theoretical side will be kept to a minimum and will be referred to only when it's necessary to clarify a particular point.

Some of the pieces of equipment used or described in this book may seem rather sophisticated for the amateur, but they are a good investment and can save money in the long run. Take, for instance, machines for grinding the head ie for re-profiling chambers and ports. The cost of these ranges from £50 upwards. One would only have to modify three or four heads either for oneself or one's friends, to more than recoup the outlay on the equipment. Those readers who are involved in any form of competitive motoring will derive great benefit from a comprehensively equipped workshop not only from the point of view of doing heads, but also in tackling a host of other jobs not normally entertained because of lack of equipment.

In the first part of the book we will deal with the methods of modifying heads as opposed to the shapes and sizes that they are going to be modified to. The reason for doing this is that a practical book on head modifying will be rather less practical than it should be if this is not done.

In the second and third parts of the book the subjects dealt with will be the head designs to which we should work and comments on the applications and results of the head modifications.

To sum up, armed with this book plus a little equipment, the amateur can produce modified heads which are at least as good as the best professionals providing the work is done carefully and conscientiously. Having, I hope, aroused your interest, we will quickly press on with the facts you will need for the task in hand.

PART ONE

chapter one

Workshop Equipment

The most essential piece of equipment for a head modifier is a grinder which can be easily manipulated into awkward places such as half way round the curve of a port. The cheapest way to achieve such a set-up is to acquire a flexible shaft outfit from your nearest do-it-yourself shop and drive it by means of a drill gun. Although this set-up is cheap, it does, for our purposes, have a number of serious drawbacks.

The first of these drawbacks is one of speed. To be effective as a means of metal removal, a grindstone must be rotated such that its surface speed is at least 3,000ft/min. Since the average drill gun rotates at about 1,000rpm the minimum diameter of grindstone needed to achieve the required surface speed is about 11in and it will be obvious to one and all that an 11in dia. wheel will not go down the average cylinder head port. From experience I can say that the most used size of grinding wheel is about 1in dia.

Having established a wheel dia. we can then work out the rpm required to achieve the minimum surface speed for efficient grinding. A quick bit of mental arithmetic will reveal that the minimum rpm requirement is in the order of 12,000. It can immediately be seen that the electric drill gun and flexible shaft falls well short of the mark. Assuming a high speed drill gun is used (one with 2,500rpm available and a chuck-mounted speed multiplier) 5,000rpm is still all that is achieved. The result of using such a low speed for grinding is that the metal removal rate is vastly reduced and wheel wear is greatly increased. To quote an example, using a 1in dia. wheel and 12,000 or more revs, it takes about ten minutes to rough grind the new profile on a Mini combustion chamber. At such a speed, the wheel will last between 8 and 12 chambers before it becomes appreciably worn. If we make do with the best that can be achieved with a high speed electric drill gun and flexible shaft having 5,000rpm maximum, then the ten minutes to rough out a chamber gets nearer two hours. On top of this we can expect to use a grinding wheel per chamber.

Another factor which affects the ease of carrying out grinding operations when using a flexible shaft on a drill gun is that most

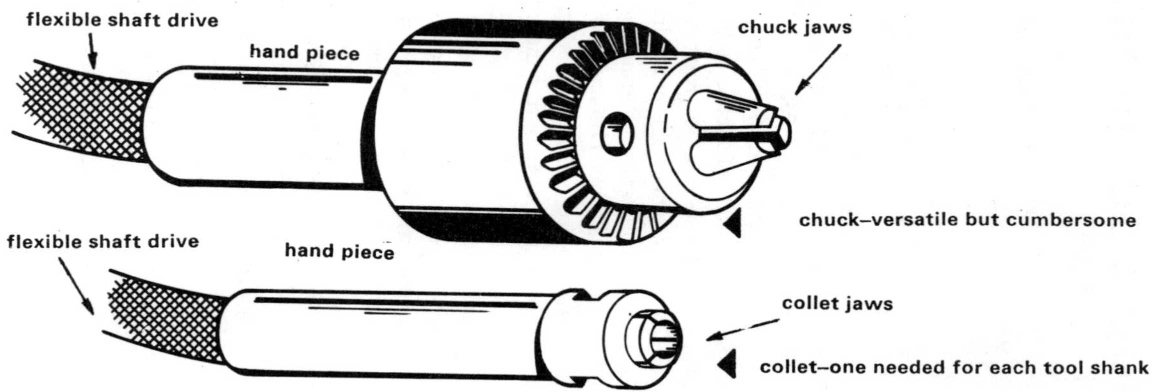

Two flexible shaft end-piece alternativcs
The chuck or the collet

proprietary flexible shafts have a chuck rather than a collet arrangement in which to hold the shank of the grindstone. A chuck is far more cumbersome to use when compared with a collet, since its relatively large diameter prevents easy access down ports. To a certain extent, this can be overcome by using a grindstone with a long shank. This allows access to the port but prevents the operator from seeing what he is doing, because the chuck is large enough to block the view.

The use of a collet in which to hold the grindstones has much in its favour although a collet is less versatile in the range of shank sizes it will accommodate. This point, however, need not concern us unduly, since grindstone shanks are invariably $\frac{1}{4}$in, 6mm, or $\frac{1}{8}$in dia., hence three collets are all that are required. The use of collets for holding the grindstones means a lighter hand piece and one which can gain access into awkward places while allowing the operator a greater chance to see what he is doing.

It seems as if we have a stack of arguments against using a drill gun and flexible shaft. How about the points in its favour? Quite honestly, this can be summed up in one word—cost. This set-up is about the cheapest way to kit yourself out for grinding heads.

Assuming you want to do the job quicker and in a more professional fashion, what alternatives present themselves? There are, fortunately, a number of firms which manufacture specialised flexible shaft grinding machines and these do, to a lesser or greater degree, fulfil our requirements for the ideal head grinding machine. The Morrisflex 300 flexible shaft grinder is typical of this type of machine. It provides sufficient rpm, up to 15,000 being available, and on top of this, adequate power, an important factor which, as yet, has not been mentioned. The power requirement for grinding is quite high. If insufficient power is available, especially when high revs are being used, the grinder will stall as soon as the grinding wheel touches the workpiece. Assuming that speeds are in excess of 10,000rpm, then the minimum power requirement for the motor in our particular case is between $\frac{1}{5}$ and $\frac{1}{4}$hp. Apart from the advantage of high speed and adequate power, most machines of the type in question are equipped with lightweight hand pieces and collet mounting for the grindstones, all of which is conducive to a better end product. The disadvantage of such a machine for the amateur is the cost, but the outlay is justified if you intend to do four or more heads from now until you actively cease any form of high performance motoring.

Having looked at the operating conditions required for good grinding, let us delve into the subject of shapes, sizes and types of grindstone to use, as the selection of a particular type of grindstone can make a great difference to the ease of producing the end product.

Firstly, the type of grit used in the stone dictates, to a great extent, the use to which it can be put. Mounted grindstones commonly use two types of grit, these being aluminous oxide and silicon carbide. The aluminous oxide stones are usually blue, blue/grey or white in colour, whereas silicon carbide stones are inevitably green and are often termed green grit wheels or stones. When it comes to grinding cast iron, the silicon carbide or green gritstone is the one for the job, especially the roughing operations. If you attempt to use an aluminous oxide wheel run at the correct speed for grinding cast iron, it will very rapidly glaze up and its grinding efficiency will drop about 95% Continued use of a glazed up aluminous oxide wheel on cast iron can produce very high temperatures at the point the wheel makes contact with the work. This can lead to small thermal cracks developing in the cast iron. The general rule then, for grinding cast iron is to use silicon carbide stones. There is, however, an exception to this, which is brought about by using small wheels at a speed which is under their correct grinding speed. With the possible exception of high speed air grinders and the like, it is very difficult to find equipment which will spin stones of $\frac{3}{8}$in–$\frac{1}{2}$in dia. at high enough revs to give the correct grinding speeds—a $\frac{1}{2}$in dia. wheel needs 30,000–45,000rpm. The use of a $\frac{3}{8}$in–$\frac{1}{2}$in dia. green grit wheel at speeds of 10–15,000rpm gives a rapid rate of wheel wear. The slower a stone is run, the more rapidly it wears. It has already been pointed out that aluminous oxide wheels run at the correct speed, glaze up when used to grind cast iron. This is because it is too hard and does not break down or wear fast enough to expose fresh grinding grit to the work when the top layer of grit has been blunted. The answer to the dilemma of a small green grit wheel wearing too fast and a blue grit—aluminous oxide—wheel being too hard, is to (a) get the speed high enough to use a green grit wheel or (b) which is far simpler, use a small blue grit wheel below its normal grinding speed and this will prevent it glazing up on cast iron.

To sum up all that has been said so far: when grinding speeds are between 10–15,000rpm we should use green grit wheels, when the wheel dia. is between $\frac{5}{8}$in and 1in, and blue grit wheels when the dia. is less than $\frac{5}{8}$in. If revs available are appreciably below 10,000, the best results will be had by using blue grit stone for all sizes up to 1in dia.

Before we move on to the shape and size of wheels to use, a little needs to be said on the dressing of a grindstone. The way a grinding wheel is dressed influences both its cutting rate and the finish it gives. If you are going to rough out the ports or chambers in a cylinder head, then the metal removal rate is more important than the finish. To get

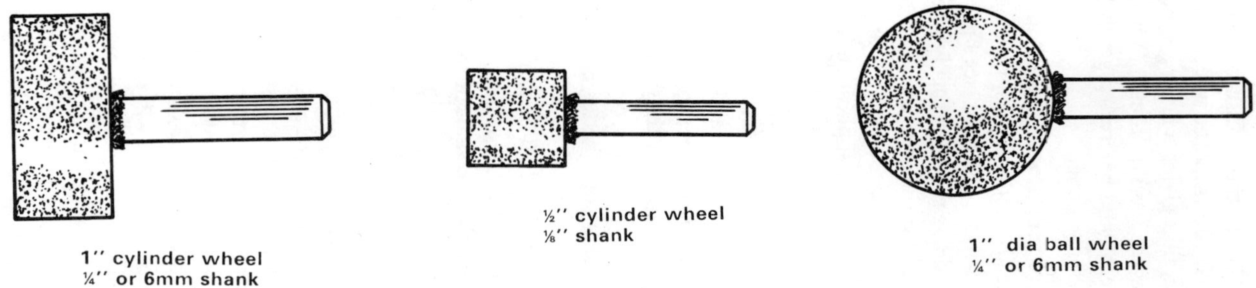

Fig 1 Grinding stones

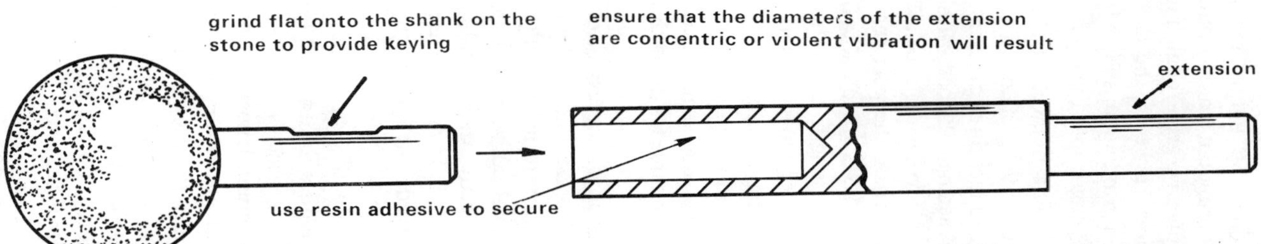

Fig 2 Method of extending shank length

the wheel to cut at its maximum rate, it should be rough dressed. This is done by lightly tapping the wheel with the dressing stone—carborundum stone—while it is running. By dressing the wheel in this manner we get a rough finish on the stone, thus giving a faster grinding action. However, the finish produced on the work is rough. When a good grinding finish is required, the wheel should be dressed gently by bringing it into light contact with the dressing stone.

The number of shapes and sizes of grindstones needed to do virtually 99% of the cast iron cylinder heads produced is surprisingly few. In fact three different wheels are basically all you need (see Fig. 1). For instance, the ball wheel is excellent for doing ports, especially where a nice curve is required as would be the case for the area just under the valve where the port curves to the outer edge of the head. If any of these wheels are found to be too large for a particular application it does not take long to dress them down to the required size. Grindstones in general tend to have short shafts, which makes for great rigidity, a prime factor for good grinding. However there are many instances when a long shaft is needed to gain access to certain areas. The way to get a long shaft on these grindstones is to extend the existing one, Fig. 2 shows the technique for doing this. The standard shaft sizes for the grindstones are $\frac{1}{4}$in, 6mm and $\frac{1}{8}$in dia., for $\frac{1}{4}$in and 6mm dia. shafts, use a $\frac{3}{8}$in dia. material for the extension piece; $\frac{1}{8}$in dia. shafts are better off with a $\frac{1}{4}$in dia. extension. The end of the extension should be machined if necessary to fit the collet or chuck size being used. Where high revs are involved, we must watch out for two things, that is (a) the extension is perfectly straight and (b) the location hole for the shaft of the stone and the holding dia. are perfectly concentric. If this precaution is not observed, the grindstone can burst and the extension bend itself through 90°.

By this stage the reader should be well acquainted with the techniques for grinding cast iron heads. However, there are numerous cylinder heads fitted to production cars which are not cast iron but aluminium. Under no circumstances should you attempt to grind heads which are made of aluminium. The roughing out of these heads requires the use of cutters made of either high speed steel or tungsten carbide. As with the grindstones, a few shapes of cutters will serve most purposes. The shapes shown in Fig. 3 will, ninety-nine times out of a hundred, do the job required. With aluminium, a fairly coarse-tooth cutter is required. If a fine-tooth cutter is used it will rapidly clog up and cease to cut. When using these rotary cutters on aluminium, it is advisable to keep them well lubricated with paraffin. By doing so, the chances of the cutter clogging is reduced almost to nil, a better finish is

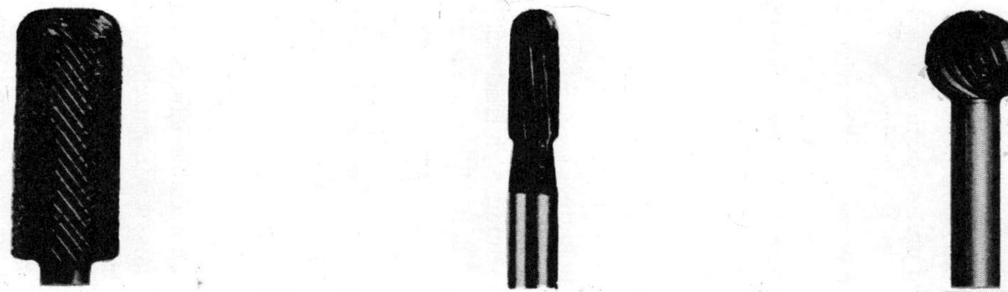

Fig 3 These three shapes of high speed steel cutters are sufficient for most jobs on aluminium heads

Tungsten carbide cutters, more expensive but wear life and cutting rate are considerably improved

obtained, the cutter lasts longer and the metal removal rate is higher. Cutters can be run at similar speeds to grindstones, especially when cutting aluminium. They are, however, far less sesnsitive to speed as far as efficient cutting is concerned than are grindstones. As mentioned previously, these cutters can be made of high speed steel (HSS) or tunsten carbide. The HSS cutter performs a satisfactory job, but does not last as well as the tungsten carbide items. Tungsten carbide cutters will outlast HSS cutters by a factor of as much as twenty when used on aluminium. The only snag is that the carbide cutters are about five times the price, therefore the extra cost isn't justified unless you intend to do a large number of aluminium heads.

The final polishing of a head casting is the operation that seems to defeat most people. Fortunately it is one of those jobs which, if you have the right bits and pieces and apply the right technique, becomes very simple and relatively quick. Equipped with the right gear, a head can be fully polished in less than an hour. The finishing or final shaping and rough polishing operations are combined.

The first task is to finish and rough polish ports and chambers with coarse emery drums. These emery drums are fitted on to special rubber mountings as depicted in Fig. 4. The mountings are available in three shapes to satisfy differing requirements. Round nose ones are best for doing ports. Since the shaft is short on these mounts, it is advisable to extend them to allow complete access to the ports. Since the round nose type is the best for ports, it is best to extend this one and leave the others with the standard shaft length. The square nose type will be found most useful for polishing chambers. Where chamber walls are straight, the square nose emery drum is best mounted on its correct square nose mounting, but if chamber walls vary from straight to curved eg Cooper S, they are best fitted onto the cone-shaped mounting. The reason for so doing is that when the emery drum is new it will follow the straight faces of the chamber walls without appreciable deformation. As the drum wears, it becomes more pliable, thus blending itself to the curved part of chamber walls and smoothing them out without destroying the curve. When using a square nosed emery drum whether square or conically mounted one should take care as they are liable to detach themselves. This can be dangerous especially if the eyes are not covered so wear goggles. Once the rough polishing has been done, we can switch to fine grade emery drums which will give us a very smooth surface finish. The fine drums are used in exactly the same manner as the coarse ones, so nothing further need be said about them.

The use of these emery drums does not, in certain instances, provide the complete answer to polishing and blending of ports. Take

Fig 4 Fitting emery drums and some that are available (Photos courtesy of Morrisflex)

for example an 850 Mini head, where each of the two ports splits up to feed two inlet valves. There is a point half way round the bend in the port where it is extremely difficult to achieve a continuous curve. Fig. 5 shows the profile normally obtained by using emery drums only in the finishing and polishing stages. To eliminate the kink in the port, we need to make up a special kink eliminator as shown in Fig. 6. This particular little tool is extremely useful as it has a very strong blending action when new. As it begins to wear, it becomes a very good polisher and so kills two birds with one stone. A look at the drawing of this tool will reveal that it is made up of a number of adhesive backed abrasive discs. A number of these discs are drilled at a very slow drilling speed to provide the central hole. The drilling of a number of the discs at one go is made easy by the fact that they come in packs of fifty all stuck together. Once the pack has been drilled, discs are removed from the pack and stuck back to back, ie adhesive side on adhesive side. The discs are then mounted on the spindle as shown in Fig. 6 with a washer between each pair of discs. The whole lot is then secured by the nut at the end of the spindle. To use this blending tool, one simply adopts a back and forth action, whilst applying light pressure over the area being blended.

The roof of a combustion chamber must be treated in different fashion from chamber walls or ports. Both the blending tool and emery drums are ineffectual as far as finishing and polishing the roof of a chamber is concerned. For treating the roof of a chamber, we must use abrasive discs in the manner for which they were originally intended. The abrasive discs are stuck on to a rubber mounting (Fig. 7) thus enabling a flat surface to be ground and polished. The use of very coarse grit discs on the roof of chambers seems to be a waste of time. Remember that for port polishing a similar speed will be required as that for grinding.

As mentioned previously, we cannot grind aluminium heads with grindstones but when it comes to finishing and polishing, we can adopt an almost identical technique to that used for cast iron heads. The only snag with aluminium is that it causes the emery drums, especially the coarse ones, to wear must faster than they do on cast iron. The rate of consumption of coarse emery drums is increased by a factor of between two to three over the requirements of cast iron. The fine emery drums, when used on aluminium, have a wear life nearly the same as when used on cast iron. The blending tool as shown in Fig. 6 can be made to give a relatively high polish if it is used with a mixture of metal polish, oil and paraffin in approximately equal proportions.

There is a popular misconception that all you need to polish

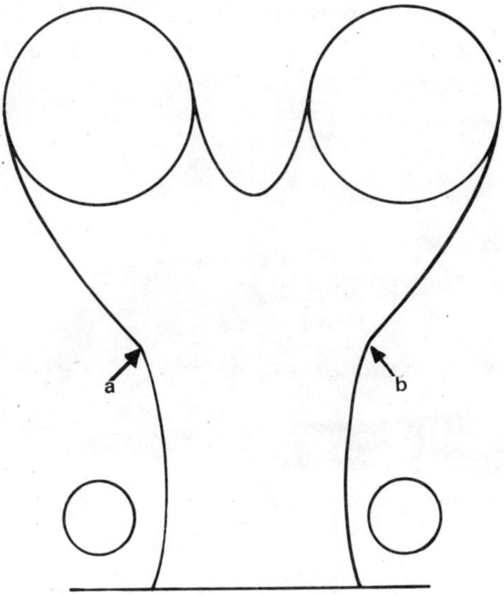

a and b are points of discontinuous curves.
a kink eliminator is needed to produce a
continuous curve at these points

Fig 5 BLMC (BMC) series A inlet port

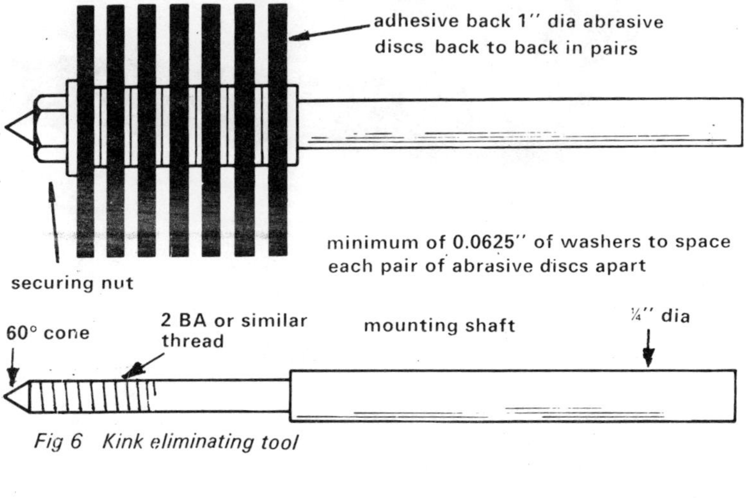

Fig 6 Kink eliminating tool

ports is a strip of emery wrapped around a bar of metal. This method of polishing is highly inefficient when compared with the methods which have just been discussed.

Fig 7 A selection of typical grinding wheels

chapter two

Valves, Guides and Seats

Before we can commence with the modification of a cylinder head, it will be found necessary to completely strip the casting of all fitments such as studs, valves, guides, etc. The first problem arises when the studs are due for removal. The simple method of removing studs (Fig. 8) is usually satisfactory if none of the studs are rusted in. In practice, one normally finds that half the studs are, in fact, corroded and the method of removal shown in Fig. 8 is inadequate. The answer, then, is to acquire a proper stud extractor. These are relatively inexpensive and definitely save time, trouble and frayed tempers.

Stripping down the rest of the head is straightforward enough until we come to remove the valve guides. The correct way to remove valve guides is to have them pressed out. This involves the use of the sort of equipment that your local motor machine shop will have, so you can easily get the job done there. If you want to do the job yourself, then make up, or get made up, a punch or drift as shown in Fig. 9. With the head properly supported, the drift can be used in conjunction with a large hammer to knock the guides out.

When guide refitting time comes, after the head has been modified, remember never to use old guides. Always fit new ones. The reasons for this are two-fold (a) the tool for cutting valve seats has to locate on the bore of the guide and it depends on accurate location to produce an accurate seat and (b) worn valve guides cost power. Due to small eccentricities which exist between the outer dia. and the bore of a valve guide, it will be found that the new valve seat, when cut, is in a slightly different position from the original valve seat. The effect of this can easily be seen whilst the seat is being cut. The seat cutter starts to cut on one side of the original seat but not the other. By the time the seat has been cut over a full 360°, we find that we have an uneven seat width (Fig. 10). For this reason we must leave the finishing of the port at and just below the valve seat until after the seats have been cut. If things are done in the reverse order, we can end up, in extreme cases, with the valve seat of the valve completely mismatching the seat in the head over part of its circumference. The golden rule, then, is to cut seats with new

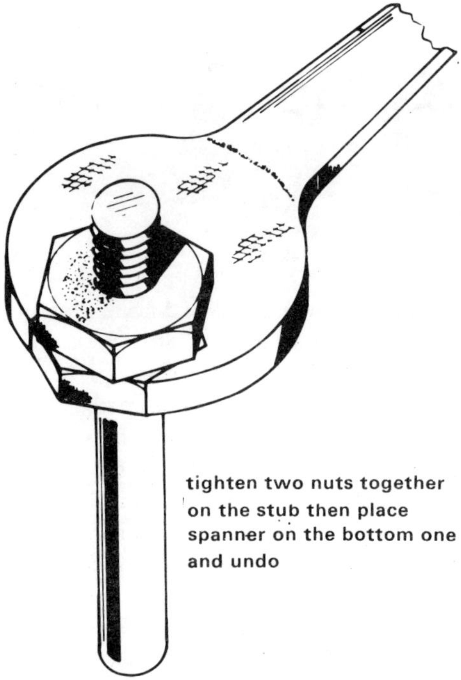

Fig 8 Removing tight stud

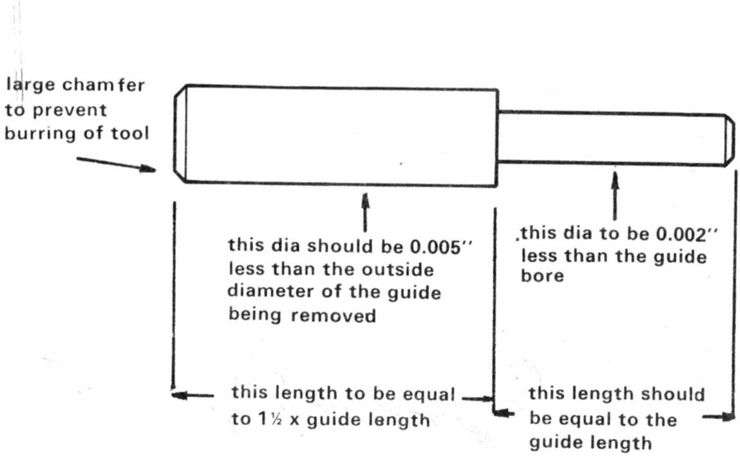

Fig 9 Guide removing punch

guides fitted. Once the seats have been cut, don't remove them but do all the final trimming of the seat width as a last operation.

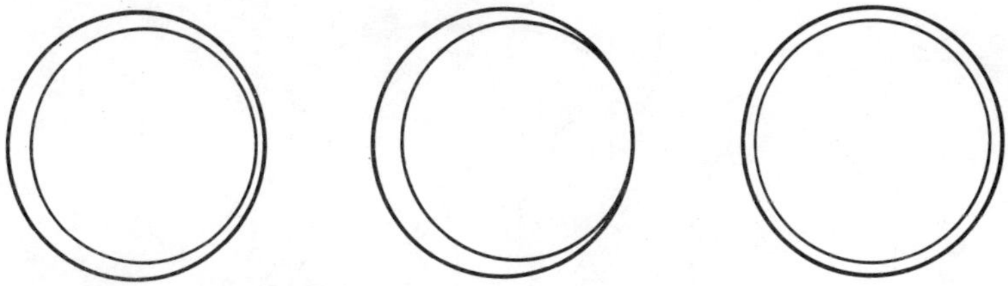

Fig 10 *Left: typical mismatching of throat dia to valve seat; centre: no seat over part of throat diameter; right: the desired result of trueing up a seat such as on left*

At this point, it may seem that the cutting of a conventional 30° or 45° seat has been dealt with lightly because it is a simple operation. Even though it is a simple operation there are, for our purposes, good and indifferent methods of producing a seat in the head. The most commonly used method to cut a seat is by a vibrating grindstone which has been previously dressed at the correct angle. The grindstone is located by a pilot in the valve guide and then driven through a spring arrangement by a pistol drill. When driven, the spring sets up a torsional vibration to assist accurate grinding of the seat form. From our point of view, this is the least satisfactory method of producing a valve seat because (a) we do not have very precise control over the seat width (b) the seat produced is a little on the rough side, and (c) all the loose grit from the stone can lead to wear on the bore of the guide. Some of this type of grinder do, in fact, use fixed pilots in the guides which does, of course, invalidate the last point, but the other two still apply. The necessity for frequent wheel dressing is another point against this type of seat finishing equipment.

Whilst the last method of grinding seats has its limitations, a grinding operation, if done in a sophisticated enough manner, need not be condemned. There are machines available which do a really excellent job of valve seats, these are of the orbiting valve seat grinding type of machine. The machine is first located concentric with the guide by a precision expanding mandrel. This ensures a dead fit thus eradicating the effects of size tolerance on the bore of the guide. The grinding machine is then located on the mandrel in a similar manner. Once the grinder has been located and clamped down, the mandrel is removed.

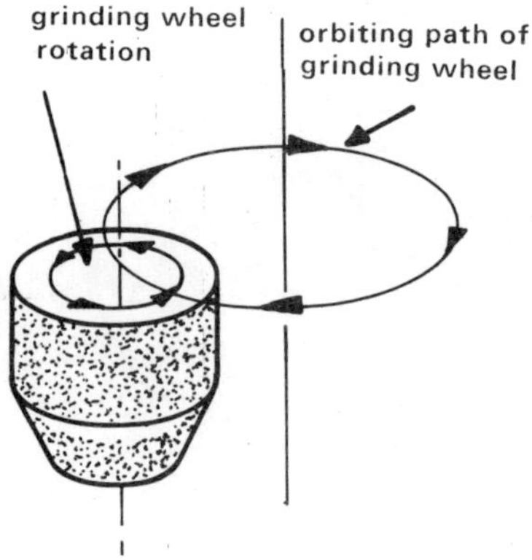

Fig 11 Relative motions of an orbital seat grinding wheel

The grinding wheel is driven at high speed (in the order of 10,000rpm) to give the correct surface speed to the wheel, then the complete wheel head is rotated in order to cut the seat over its entire circumference. The diagrammatic layout in Fig. 11 will probably clarify the exact motion of the wheel. Because of the expense involved in buying a machine such as this, very few engine machine shops seem to possess one, but this situation is changing for the better.

Apart from grinding seats by using the aforementioned equipment, we can use cutters. By using good seat cutters we can obtain a seat which is very nearly as accurate as the orbital seat grinding machine. However, the type of cutters used does influence this factor. The best, and most expensive, seat cutters use adjustable tungsten carbide blades or inserts to provide a cutting edge. These produce a high quality seat even if the seat is glazed through prolonged use. Less expensive high speed steel or the even cheaper high carbon steel cutters can be blunted very rapidly if they are used to cut glazed seats too often. To quote an example, high carbon steel cutters can be blunted by cutting as few as eight seats, and high speed steel ones as few as forty to fifty seats before becoming blunt. It is these cheaper seat cutters that the do-it-yourself head modifier is likely to acquire. To get a reasonable life from these cutters, certain precautions should be observed. The first of these is obviously in connection with the glazing of the seat in a used head. Before removing the old guides, it is a good idea to lap in the valves with coarse grinding paste to remove the glazing from the seats. If this is done, the cutter life is extended by a factor of about ten, and the accuracy of the final seat is better, since when the time comes to finally cut the seat, far less pressure is required.

Since we have got around to the subject of pressure required to cut a seat, this is as good a time as any to enlarge upon the subject. To get an accurate seat, it is essential that only light hand pressure is used. If heavy pressure is required to cut a seat, due to cutter bluntness or glazing, the cutter can be biased from the true centre line of the valve guide. The final result, after an extended blister forming lapping session, is an untidy, unworkmanlike job. If all the precautions outlined are observed, cheaper seat cutters will perform satisfactorily.

Conventional 45° angled seats are adequate for road going engines where life, as well as power output, must be considered. For a racing engine, where ultimate life is of far less consideration, we can afford to treat the subject of valve seats far more technically. The power output of an engine is greatly affected by the amount of air it can draw in. If this is increased, the power will go up. It is a well known fact that radius valve seats greatly increase the discharge coefficient of a poppet valve. To put this into simple terms, the efficiency with which a valve can pass air is increased if it has a radius seat instead of a normal 45° seat. To produce a good radius valve seat with limited equipment is a little more difficult and a lot more time consuming but it certainly is one step nearer to gaining the ultimate in power from an engine.

To do radius valve seats on a particular head, you will need a tool as shown in Fig. 12. The actual tool bit is best made up from a broken or worn out centre drill. Do not attempt to make it up from the shank of an ordinary drill as these are not hard. If you haven't got a worn or broken centre drill, then a new one will do the trick just as well. They only cost a few shillings and can be obtained from any engineer's tool shop. The problem is, in fact, putting on the required radius to form the seat. If the following technique is adopted, the problems are diminished to the smallest magnitude possible short of giving the job to someone else to do. The first move is to obtain a piece of brass bar or failing that, a mild steel bar, and have it turned to the correct size. The size required for the bar will be between 0.2 and 0.3 times the diameter of the valve. The radius we will produce will therefore be between 10% and 15% of the valve diameter. Using the brass or steel bar as a radius gauge, rough out the radius on an offhand tool grinder. It is best to get the roughed out radius as close as possible to the required radius, as the next operation can be long-winded if there is a great discrepancy. The next step is to mount the brass or steel bar in a lathe or drill gun. If a drill gun is going to be used, hold the handle in a vice to steady it. Liberally apply fine grinding paste to the bar and start the lathe or drill gun. Place the roughed out radius onto the bar and with

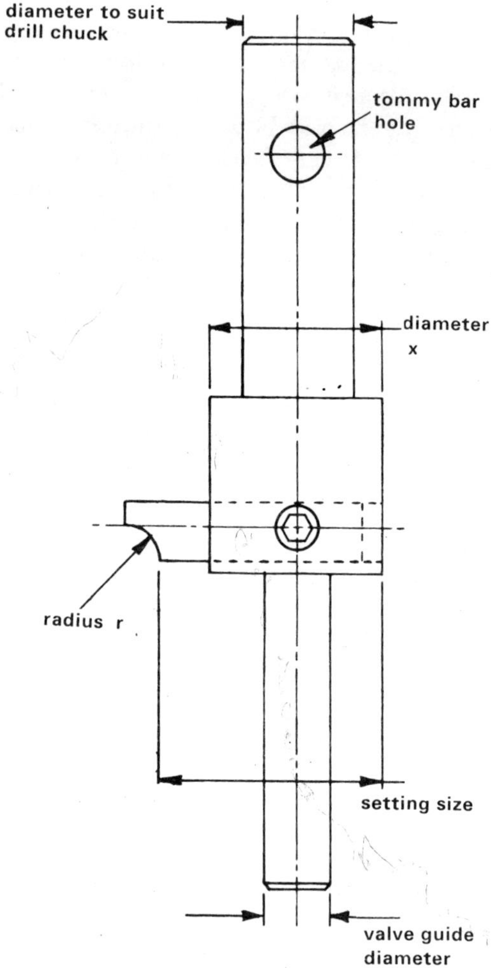

$$\text{setting size} = \frac{\text{valve dia}}{2} - 0.293r - 0.020'' + \tfrac{1}{2}x$$

$$\text{where } r = \frac{\text{valve diameter}}{10}$$

Fig 12 Radius valve seat cutter

sweeping movements, move it up and down the bar. The reason for this motion is to distribute any wear which may take place on the bar. After about twenty minutes you will find that the bar has reproduced its form accurately onto the tool bit. When producing the radius it is essential that we angle slightly less than 90° to the bar so as to produce a cutting clearance (Fig. 13).

Unless you have a lathe or access to a lathe, the tool holder can present a little more of a problem. If you do experience any difficulties, then a motor machine shop or subcontract machine shop should be able to do the job for about £2.

Once the cutter has been made up, we can get down to producing the valve seats. In most cases, to use radius valve seats we need to go up in valve size. This is because there is insufficient metal under the valve seat to allow us to produce the full radius (see Fig. 14). The machining of the seats should not be attempted until the head is finished as far as ports and chambers are concerned. The procedure is to grind, profile and polish ports so that they are finished except for the last $\frac{1}{4}$in or so immediately under the valve seat.

The chambers should also be ground to shape and finish polished. At this point the valve guides can be fitted and a rough cut done on the seats. The roughing cut should be 0.010in less than the final cut which, of course, can be worked out from the formula in Fig. 12. Once the roughing cut on the valve seats has been done, the ports and chambers can be blended into the seats. Once everything is looking approximately as it should, the seat cutter can be set to its final size. The final cut should be done on a pillar drill and the cutter should not exceed 100rpm. If too high an rpm is used the seat will be marred by chatter marks which can be very difficult to get rid of. The ideal speed to make sure chatter does not occur is around 40–50rpm. Once the seats have been finish cut, they have to be checked for accuracy and trued up, should any errors of truth be found. Even if great care has been taken, it is possible to have the seat eccentric by two or three thou with respect to the valve guide, so steps must be taken to eliminate such errors. To get the seats true within the required limits, take a valve, apply some fine lapping paste to the seat and lightly lap it on to the seat of the head for about a minute. Remove the valve and thoroughly wipe off all traces of grinding paste, then apply a thin smear of engineers' blue to the entire seat of the valve. Drop it back into the seat in the head and rotate it once or twice. When the valve is removed the blue will show the point or area of contact that is made between the valve seat and head seat. If a continuous blue line is shown over the entire 360° of the head valve seat, then the seat is concentric with the guide. If, as is often

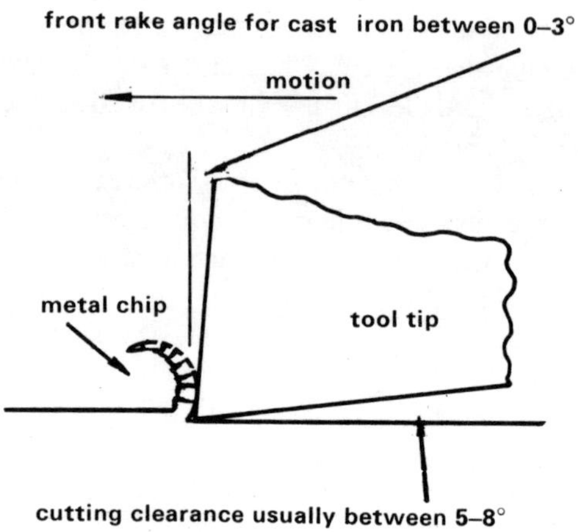

Fig 13 Cutting clearance

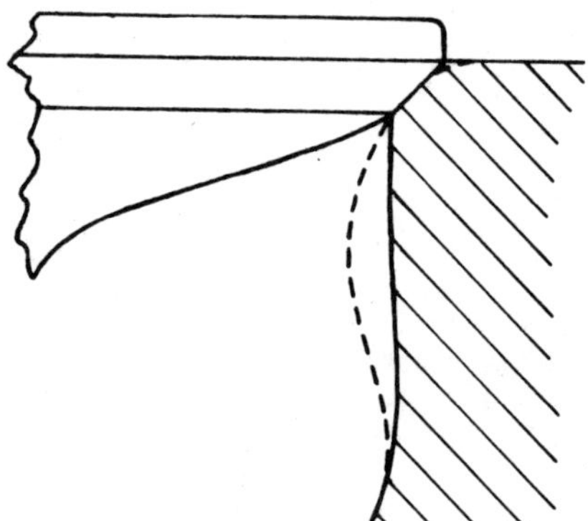

Fig 14

Insufficient metal under valve seat to produce a full radius. Solution is to fit larger valve or valve inserts

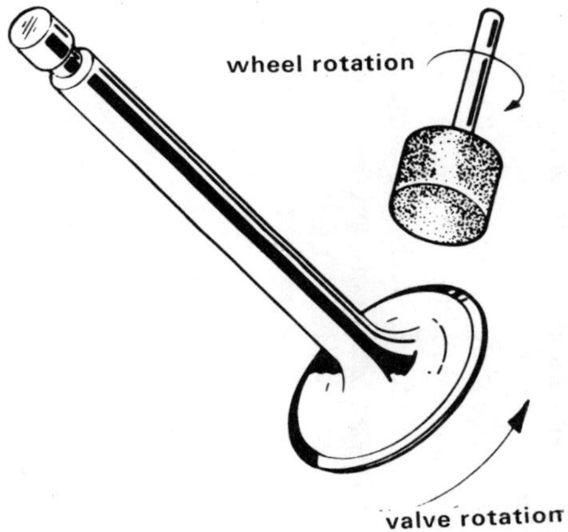

Fig 15A Profiling the under-head area of the valve

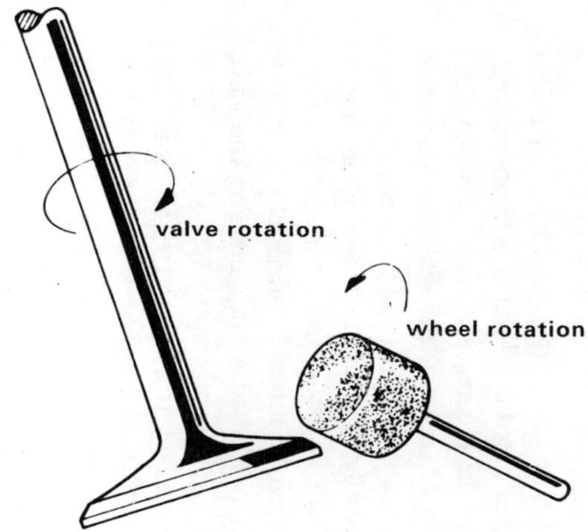

Fig 15B Blending the under-head area into the valve seat

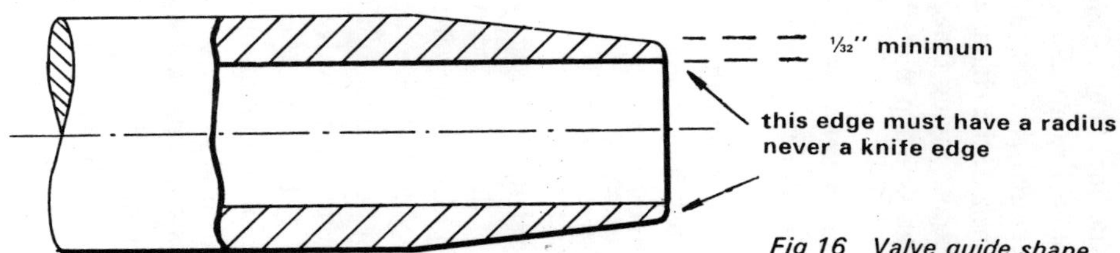

Fig 16 Valve guide shape

the case, the blue has only marked up part of the seat, it is obvious that there is undue error in concentricity between seat and guide. Take the radius seat cutter, still at its finishing size setting, and by hand lightly recut the part of the seat which has blued up. After this has been done, lightly lap in the valve and repeat the blueing procedure. Continue the lapping, blueing and cutting procedure until the seat shows an even blue line about 0.030in wide around the full 360° of the seat. This may sound like a tedious job but usually two or three goes is all that is required and only takes about half an hour. Once the seat has been finished, the chamber and port can be finish blended into it. It goes without saying that great care is needed to avoid damaging the valve seat. One slip and a lot of careful work can be spoilt.

Just as seats affect air flow, so does the shape of a valve and, to a lesser extent, so does the shape of a guide or at least the part that protrudes into the port. Reshaping a valve is simple enough if you have a lathe in which to rotate it. If not, the job can be done with two pistol drills. Mount the valve to be reshaped in a lathe or pistol drill; in the latter case, steady the drill in a vice. Then, using a drill or a grinder, grind the valve to the required profile while is is being rotated, (Fig. 15A). Take great care to avoid damaging the valve seat. When thinning down the valve seat and blending the underside of the valve head into the seat, you will find that there is less chance of causing seat damage if the grinding wheel to valve relationship is as shown in Fig. 15B. If the valves are not new but are fit for further service, it is a good idea to have the seats refaced. After this it will be necessary to do a little blending of the seats to form smooth curves between them and the rest of the valve form.

Inlet guides create less turbulence if they are tapered. This is a simple operation on a lathe. One thing we must avoid especially with cast iron guides is the production of a knife edge at the end of the guide (Fig. 16). The metal at the thin edge can break away and cause damage whilst passing through the engine. To prevent this, the thickness of the metal should not go below $\frac{1}{32}$in and the end of the guide should have a radius on it.

chapter three

Matching Ports to Manifolds

Port matching is often looked upon by the diy head modifier as a most difficult operation when, in fact, this is not the case.

The simplest way to match the ports of a head to a manifold is to use a manifold gasket as a template. To use the gasket as a template we need to locate it reasonably accurately on both the head and manifold being matched. As far as the head is concerned, the studs can be used for this purpose. In most cases, however, the gasket will be found to be a sloppy fit on the studs. This can be remedied in two ways, either by winding masking tape round two of the studs until the gasket is a good fit, or to make up two metal rings which will fit over the studs to serve the same purpose.

Once the gasket is located, we merely mark around the port apertures and grind them out to the lines thus scribed. To make the scribe lines show more clearly, it is advisable to apply to the cleaned faces of both head and manifold some marking blue. The procedure for marking out the manifold follows the same sequence as for the head with the exception that we may have to use nuts and bolts instead of studs to position the gasket on the manifold. The accuracy of this method of port matching is obviously limited to the amount of sloppiness of the manifold on its locating studs. If we wish to achieve more accurate matching some vast improvements in the location of the manifold to the head will be required.

To achieve a very fine matching of ports it will be necessary to dowel the head and manifold together. Dowelling these two components together will mean that we have an accurate location of the manifold on the head. To do a really accurate port matching job, we shall first require some $\frac{1}{8}$in dia. dowel pins. These can easily be cut from a length of $\frac{1}{8}$in dia. silver steel which, because of its accurate sizing, is eminently suitable for the job. From the silver steel make up some dowel pins $\frac{3}{8}$in-$\frac{7}{16}$in long and radius or chamfer the ends. The number of dowel pins required will be two for each component part of the manifold. For instance standard Mini manifolds, which have a combined inlet and exhaust, will need two dowels. A Mini Cooper would, with its separate inlet and exhaust manifold, need four dowels.

The next step is to fit the manifold studs in the head and bolt the manifold in place with two or three gaskets between the head and the manifold. The reason for the extra gaskets is that sooner or later a spare will be required and it is very difficult at a later stage, to produce a replica of the first one if more than one is not done at this point. Before tightening down the manifold, check that all gaskets are aligned or each gasket will not be the same as its neighbour. With the manifold tightened down, drill two holes as far apart as possible through the manifold and into the head. Normally the best place to drill such holes is about $\frac{1}{4}$in away from the fixing studs. The size of drill to use is 0.110in dia. and the drilling should go about $\frac{3}{8}$in into the head casting. Also make sure that the holes are drilled vertically to the head/manifold face or it will be found difficult, if not impossible, to fit or remove the manifold. Next, using a worn $\frac{1}{8}$in reamer ie one that is about 0.001in undersize, ream through the manifold into the head. Incidently, it may be found on occasions that the drill or reamer length is insufficient to reach the base of the manifold before the drill chuck fouls the manifold at some other point. This is found to be the case with Weber manifolds and the situation can usually be overcome by holding the drill in a pin chuck and holding the pin chuck in the drill chuck.

One point that should be mentioned here is that some reamers have a shallow taper at the cutting end. This type of reamer is a hand reamer. If this type of reamer is used, it is advisable to grind off the first $\frac{1}{8}$in-$\frac{3}{16}$in of its length otherwise the dowel pin will not fit because it is going into a tapered hole. The type of reamer that is most suitable is a machine reamer for it is parallel over its entire length except for a very small 45° lead in on the cutting end.

Once the dowel holes have been reamed with the undersize reamer, we can remove the manifold from the head. The dowel pins should then be fitted into the head and if the reamer used was undersize by 0.001in, they will be a light drive fit. Using a new $\frac{1}{8}$in reamer, we can now ream the holes in the manifold to produce a slide fit on the dowels. What we have now produced is an accurate location for both manifold and gasket on the head. Using the dowelling points as a location we can accurately mark out both the head and manifold to the gasket to get extremely good port matching.

There are instances when port matching by the dowelling method is awkward if not impossible. Such a difficulty can arise when an attempt is made to dowel a two-piece 40 or 45 DCOE Weber manifold on to a 1500 Ford engine. The carburetter mounting flange completely obstructs the manifold to head mounting face as far as access with a drilling machine is concerned. The only way out of this situation is to get a location ring fitted as shown in Fig. 17. A motor machine shop may be able to do the job but you are more likely to get it done at a subcontract engineering machine shop. An average sort of price for this sort of job is £6 to £7 10s so it's not too cheap and of course it can only be done when the ports being matched are round.

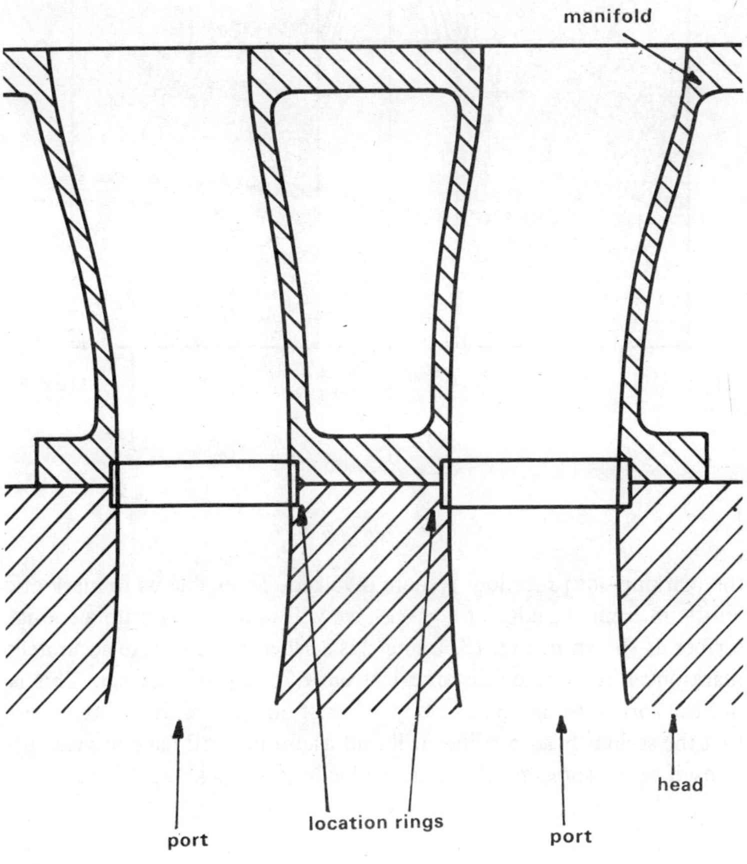

Fig 17 Matching and location of a manifold by the ring method

To come back to the subject of a gasket being used as a template for the purposes of marking out ports, we find that this is all right providing the port size required is the size of the hole in the gasket. There are, however, a number of instances when the correct port size is appreciably less than the size of the hole in the gasket. This means that

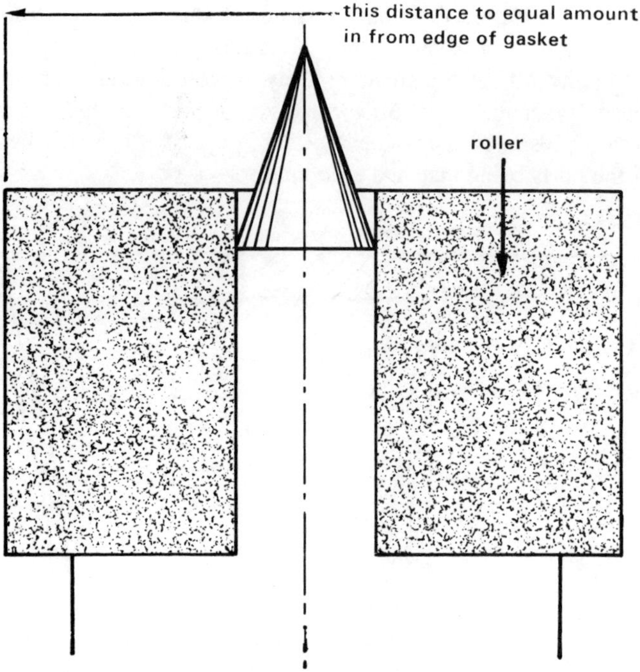

Fig 18 Roller-scriber

the marking out procedure is more involved. To enable us to mark out a little in from the edge of a gasket, we will need to make up a special scriber as shown in Fig. 18. Using this scriber we can quite accurately mark out a line a fixed distance in from the edge of a gasket. This is not the sort of thing you can buy, it's just another of those little tools that the serious head modifier will find useful but will have to make up himself or get someone else with the facilities to make up for him.

chapter four

Cylinder Head Machining Operations

There are certain jobs on a head which it is impractical for the do-it-yourself man to do. One of these jobs is the fitting of valve inserts, especially on aluminium heads. Inserts are usually required on cast iron heads if the valves are pocketed (Fig. 19) and the standard valve sizes are being retained. With cast iron heads, if we go to a larger size of valve, any pocketing will be effectively eliminated. In the case of aluminium heads, we cannot go up much in valve size before larger inserts are required, so if the original inserts are showing signs of pocketing it is best that they are replaced. If appreciably larger valves are to be fitted in aluminium heads, it is essential that there is a corresponding increase in the size of the inserts used. In some cases, the use of an insert is brought about because there is insufficient material around the standard valve seat to form the type of valve seat required. A typical case is the Mk II 2-litre and 2.5-litre six cylinder Triumph engines. To produce a venturi under the valve seat together with a radius seat, it is necessary to use an insert with the required form already on it. If an insert is being used for this purpose, great care must be taken to ensure that the insert is concentric with the valve guide bore.

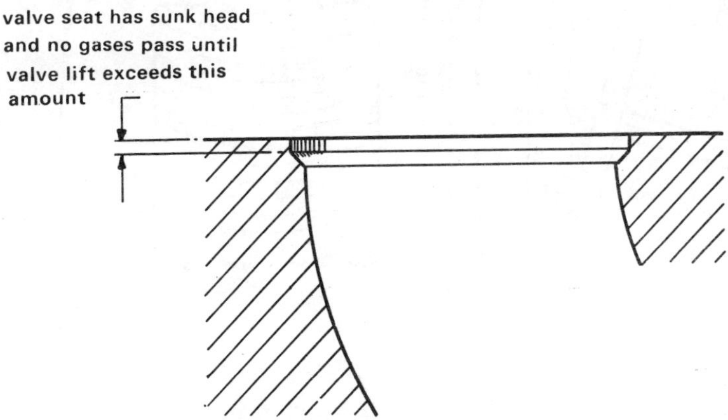

Fig 19 Pocketed valve seat

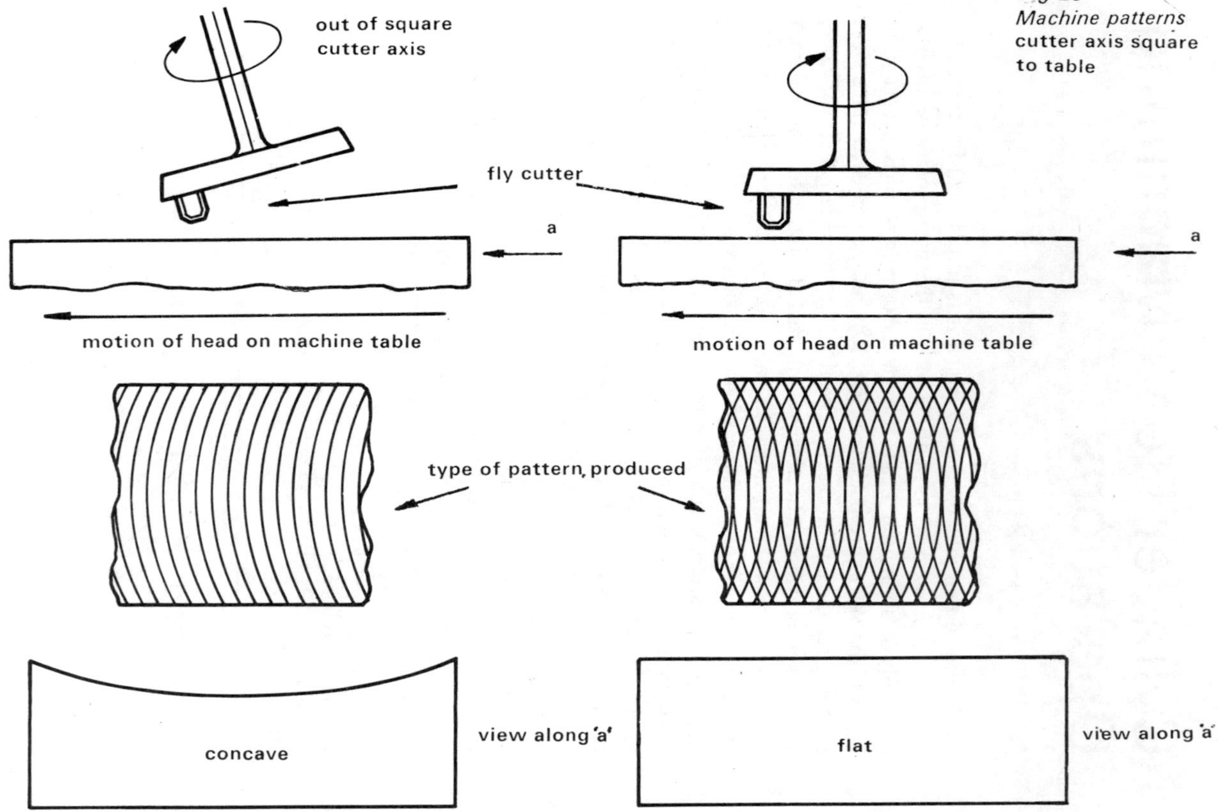

Fig 20
Machine patterns
cutter axis square to table

Head skimming is another machining operation which in all probability will not be done by the home tuner, but it is something that he should know about. There are several methods of skimming large amounts of metal from head faces, some more satisfactory than others. Fly cutting in a milling machine is a fairly popular method but if it is done incorrectly or on a worn machine it can produce a concave surface. This concave form is produced when the axis of the cutter is not exactly 90° to the motion of the machine table. If the cutter is square to the table motion it will produce a different machining pattern to one that is not square, Fig. 20 shows the difference between the two. Obviously the one produced by the cutter that is square to the table motion is the mark of a good job and is the one you want.

Heads can be skimmed on a lathe but one should be wary of having a head done thus unless the lathe used is really accurate. When a head is skimmed on a lathe, any inaccuracies in squareness on the cross slide with the lathe bed will be reproduced as twice the error on the head face being skimmed.

Yet another method of skimming is by grinding off the excess metal. Unless the concern doing the job is very well equipped on the grinding side, this process will only apply to cast iron heads. There are two common ways of grinding a head. One utilises a cup wheel, the other a normal disc wheel (Fig. 21). Of the two, the cup wheel grinding method is to be preferred. All the methods of skimming described so far are satisfactory, providing the work is done properly and with due consideration of the accuracies required.

Another machining operation which it is essential to describe is that of spring seat counterboring. Some heads, when used with a high lift camshaft, allow insufficient travel of the valve before the spring becomes coil bound. A typical example is found on the non-crossflow Ford heads. If these heads are used with a cam giving a valve lift in the region of 0.400in the coils of the spring come into contact with each other long before the full lift is reached. Once the springs have become coil bound, the valve will not open any further, so if the engine continues to rotate, something must give, the usual thing is for the pushrods to bend. If an attempt is made to run the engine while this condition exists, the cam lobes and followers will be severely damaged.

The minimum clearance between the coils of a valve spring when at full lift should be in the order of 0.010in. If it is appreciably less than this figure, the spring platforms must be machined to increase the coil to coil clearance. To find out just how much the spring platforms will need machining, the following method should be used. Firstly determine the total valve lift given by the cam being used. Just for reference, the

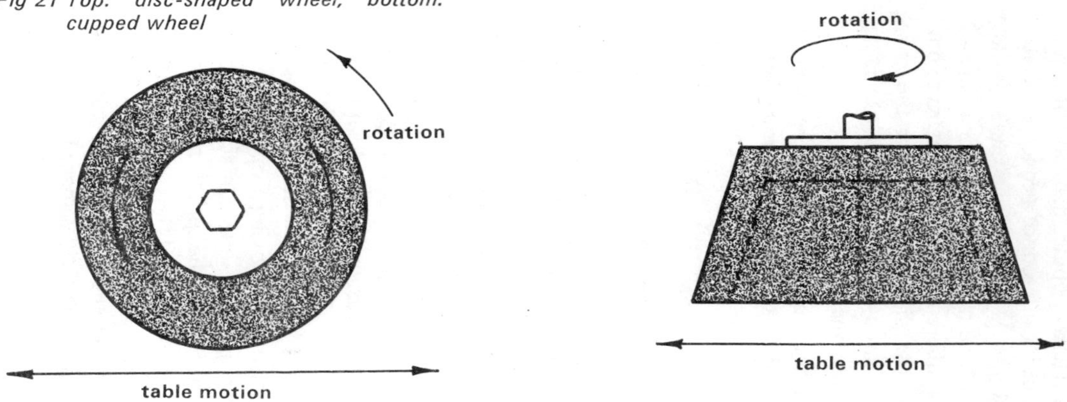

Fig 21 Top: disc-shaped wheel; bottom: cupped wheel

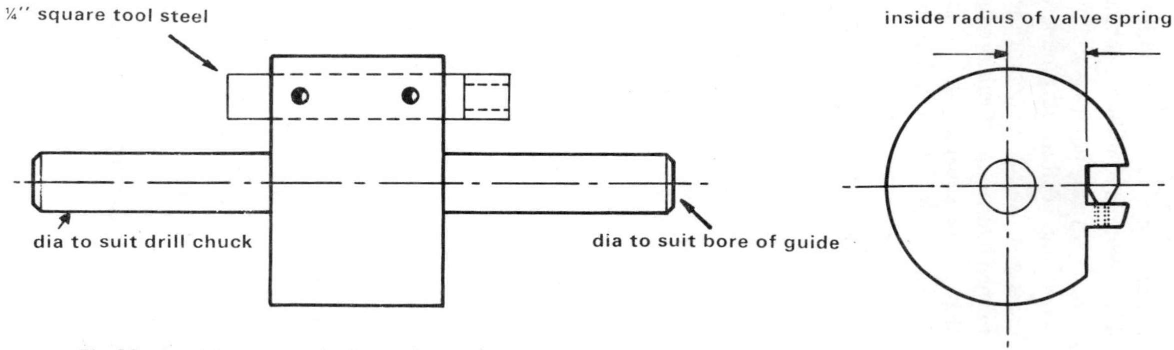

Fig 22 Tool for counterboring valve spring seats

valve lift is equal to the cam lift times the rocker ratio. Once a figure has been established for full lift, we set about measuring the maximum valve lift to give coil binding that exists at the present time. The easiest way to do this is to measure the height of the valve spring with the valve closed, then remove the valve spring, place it in a vice and close the vice until the spring is coil bound. At this point we can measure the coil bound height or length of the spring. The coil bound height of the spring subtracted from the figure for the height of the spring in the valve closed position will give us the amount of lift at present available. If we now subtract the available lift from the lift height required by the high lift cam, we are left with a figure which indicates the amount of increase in available lift required. On top of this figure we must add enough to give us the required 0.010in coil to coil clearance at full lift. To find this figure, we multiply the number of working coils by 0.010in. The resultant figure is then added to the figure for the increase in required valve lift and that's the amount to counterbore the spring platforms. To make things clearer, let's run through a hypothetical example:

Cam lift 0.250in, Rocker ratio 1.5:1

Therefore the lift at the valve = 0.250 × 1.5 = 0.375in.

Maximum valve lift to coil binding in standard form = 0.350in.

Minimum extra lift to be accommodated = 0.375in (valve lift with high lift cam) minus 0.350in = 0.025in.

Number of working coils on spring = 4.

Coil to coil clearance (Minimum) = 0.010in.

Amount of machining on spring platform = 4 × 0.010in + 0.025in = 0.065in.

Once the amount to be machined from the spring platform has been established, we can set about making up a tool to machine it. Fig. 22 gives a clear indication of what is required. If you don't want to do the job yourself, then you can get a motor machine shop to do it. The cost, however, is likely to vary quite considerably from one firm to another. The reason for this is that sometimes a valve insert boring machine can be used for the job and sometimes not, it all depends on the type of insert boring machine the firm in question has. If the insert borer is not suitable for the job of relieving spring platforms, then it is a case of having to make up a special tool as depicted in Fig. 22.

Providing the valve seats have not been cut, it is possible to press the valve guides such that they are below the level of the spring platform. Doing this will enable the valve insert borers, which utilise a pilot located in the guide, to do the job of spring platform relieving.

chapter five

CR Calculation and Chamber Balancing

With any tuned engine, the one thing that is sure to change, usually in an upward direction, is the compression ratio. Since raising the CR increases engine power output, it is desirable to have this as high as possible within certain limits. Our upper limit is always dictated by the point at which detonation occurs. Since detonation can very quickly destroy an engine, it is best that we know exactly what the CR is, or will be, so that it can be kept to sensible proportions.

The compression ratio is expressed by the following formula $\frac{V + C}{C} = CR$ where V is equal to the swept volume of the cylinder and C is equal to the volume of the combustion chamber.

To work out the swept volume or capacity of one cylinder is simple. To do so you merely divide the engine capacity by the number of cylinders ie if a four cylinder engine has a capacity of 1100cc the capacity or swept volume of one cylindor will be $\frac{1100}{4} = 275cc$. To arrive at a figure for the chamber volume is a little more complex. To establish a figure for this volume, we have to physically measure it and for this purpose we will need a pipette or burette graduated in cubic centimetres (cc). The volume of the combustion chamber is the total volume that remains above the piston when the piston is at TDC. This is made up of the volume contained in the head, the volume caused by the thickness of the gasket, the volume between the top of the piston and the top of the block, at TDC, plus the amount which may be contained in the dish of the piston if dished pistons are used, or minus the amount displaced if raised crown pistons are used.

The volume of the chamber contained in the head is measured as shown in Fig. 23. Once this has been done, we can add on the volume of the gasket. If this has a circular hole in it the simplest method is to work out the volume it will contain from the following formula:

$$Vcc = \frac{\pi D^2 \times L}{4} \div 1{,}000$$

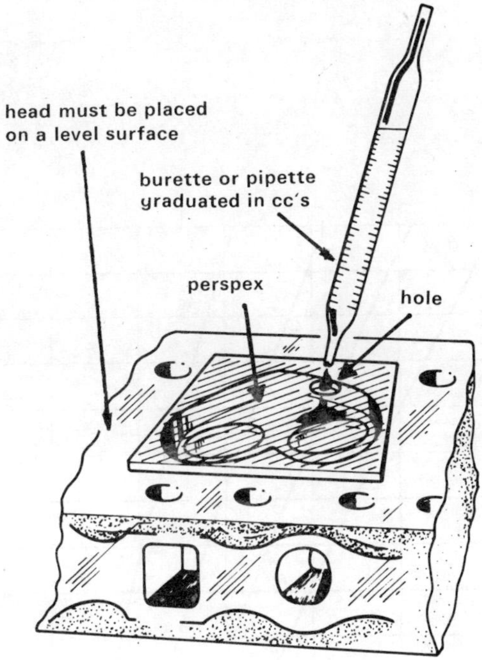

Fig 23 Measuring chamber volume

Where V = volume
π = 3.142
D = dia. of the gasket bore in mm
L = compressed thickness of the gasket in mm.

If the gasket does not have a round bore, and this is the case with many, we can resort to measuring the volume it contains with the aid of a burette. To do so, stick a compressed gasket onto a sheet of glass using gasket sealing compound, then place the glass on a horizontal surface and fill the bore of the gasket from the burette. Do not overfill the gasket bore or an entirely false measurement will result. Filling should be continued until the fluid is level with the edge of the gasket.

Since all bores are round it is easy to calculate the volume contained between the top face of the piston and the top of the block. This can be done from the formula mentioned but D will equal the cylinder bore dia. in mm and L will equal the distance from the top of the piston face to the top of the block again in mm. The graph (Fig. 24) will save time on the calculations of volume contained in gaskets, so long as they are round and can also be used to determine the volume contained in the block when the piston is at TDC.

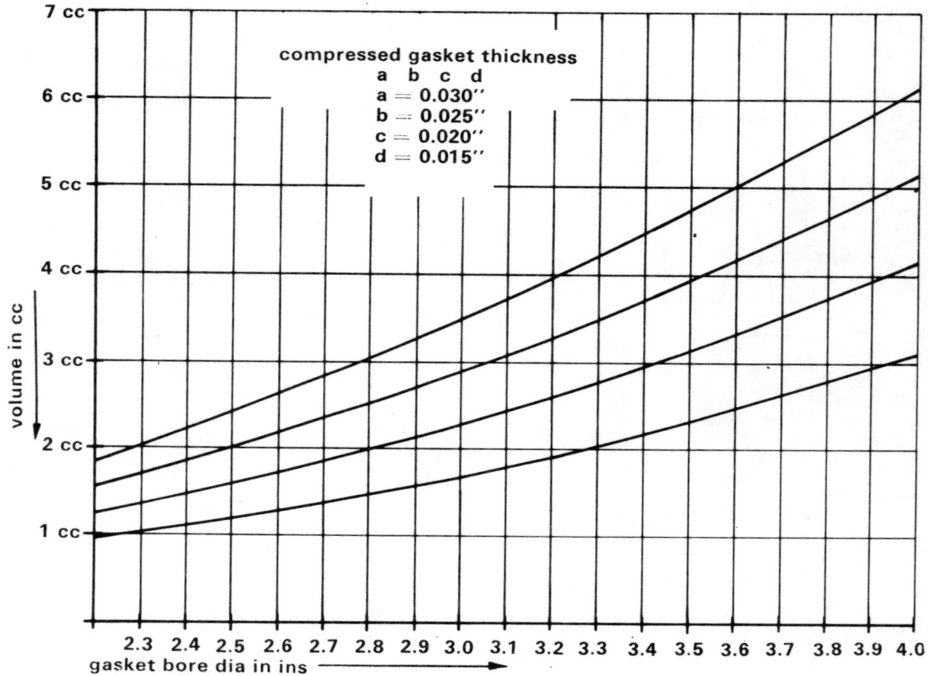

Fig 24 Determination of cylinder bore volume

At some stage it will become necessary to establish how much metal is to be removed from the face of the cylinder head to give the desired compression ratio. To do this we must first calculate the total chamber volume required. From the resultant figure we subtract the volume of the gasket, the volume contained in the block above the piston when it is at TDC and, if a dished piston is used, the volume of the dish. The figure we are now left with is the volume that the head must contain to give us the desired ratio. To make sure that's clear, we will just run through an example.

Let us assume in this instance that the CR required is 10/1 and the capacity of the engine is 1,000cc and it has four cylinders.

$$CR = \frac{V + C}{C}$$

where V is the swept volume of one cylinder and C is the total chamber volume.

Since we know V. $\left(\frac{\text{capacity}}{\text{no. of cylinders}}\right) = \frac{1{,}000\text{cc}}{4} = 250\text{cc}$ and we know the CR required, we will have to transpose the equation to give us C, the chamber volume. When this has been done, we get the following:

$$C = \frac{V}{CR - 1}$$

$$\therefore C = \frac{250}{10 - 1} = \frac{250}{9} = 27.7\text{cc}$$

Therefore the total chamber volume equals 27.7cc. From this we subtract all the parts of the chamber volume which are not in the head. Let us assume the piston is dished and the dish contains 6cc and that the amount remaining above the piston when it is at TDC and the face of the block is $1\frac{1}{2}$cc. In addition to this, the gasket contains $3\frac{1}{2}$cc. The sum total of all these volumes that are not contained in the head is 11cc. To achieve our 10/1 CR we must have a volume of 27.7 minus 11cc equals 16.7cc contained in the head. To find how much will have to come off the head, we place the head on a horizontal surface, or to be more precise, we place the head such that the head face is horizontal. Having done this, we fill the chamber with the amount of fluid equal to the finished volume that is required. In the case of our example, this will be 16.7cc. The distance from the head face to the surface of the fluid is then measured and that is the amount that must be removed. One little problem does arise when measuring the distance from the head face to the fluid. As the end of the depth gauge nears the fluid, the fluid, through capillary action, comes up to meet the depth gauge. This capillary action occurs, when using paraffin as the volume measuring medium, when the end of the depth gauge is between 8 and 12 thou from the surface of the fluid and due allowance must be made for it (Fig. 25).

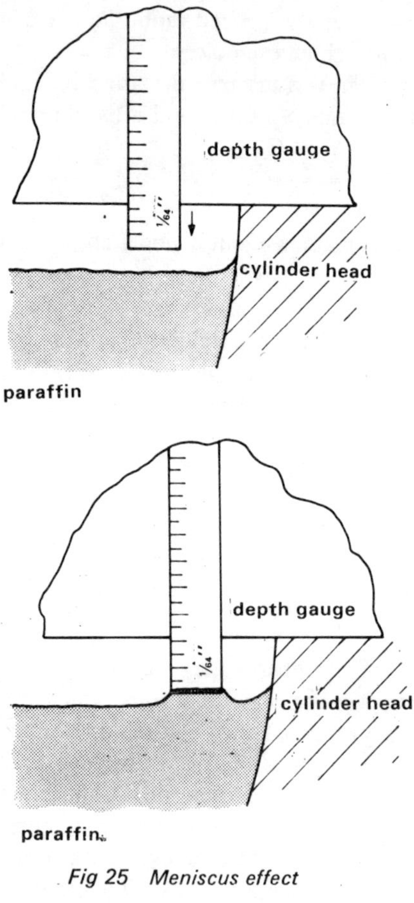

Fig 25 Meniscus effect

Because of small inaccuracies incurred during the grinding and profiling of a combustion chamber, it is advisable to check that the volume of each chamber is the same as the others. If they are not all the same, metal should be removed from the chambers containing the smallest volume to bring them up to the same volume as the largest. The primary reasons for balancing the chambers are that the engine will run smoother especially at tickover, and vibrations will be slightly reduced with even firing pulses. The second reason is that if we are using the maximum compression ratio possible, and we happen to check the chamber with the largest volume for the purposes of determining the amount to skim off, the other chambers can go above the limit. The result of this is detonation which will quickly lead to a ruined engine.

When removing metal from the chambers, it is best to take it from the roof of the chamber or from the walls adjacent to the spark plug.

The limits to which chambers can be balanced is in the order of 0.2cc. To attempt to get closer than this is a little impractical as the measuring equipment used is bordering on the limits of its own inaccuracies. Apart from this an error of 0.2cc represents, even on a small engine, a low percentage error of the total volume of the chamber in the head.

chapter six

Chamber Reshaping Methods

A great many production cylinder heads, in order to produce the maximum power, need to have the shape of the chambers changed. Both the plan shape and sectional shape are of great importance, greater in fact than many professional tuners realise. There are machines available which can be programmed or can copy three-dimensional profiles to a high degree of accuracy. Unfortunately such equipment costs many thousands of pounds so we have to make do with less sophisticated methods to produce the required shapes.

The first step to produce the required shape is to make up a template of the chamber shape's plan view (Fig. 26). This has to be located on the head in its correct position. In many cases we cannot use the existing outside shape of the chamber to align the pattern because it is a cast shape and is by no means accurate enough. There are, however, two points in any chamber upon which we can rely as far as positional accuracy is concerned, these being the valve centres. Apart from being the points of greatest accuracy, they are also the points or areas which we are either flowing gases to or from, and so their selection as location points is raised in importance. To accurately establish the plan view shape of the chamber on to the head we will need to drill two location points on the template whose centres correspond to the centre distance of the valve guides. Two valves can then be passed through these holes into the valve guides (Fig. 27) thus accurately positioning the template. It is then a simple step from here to mark around the template with a good sharp scriber.

It is a good idea prior to the marking out of the plan view shape to thoroughly clean the head face and apply marking blue. This will make any lines subsequently scribed on the head face much clearer.

The sectional shape of a chamber is a little more difficult to produce accurately since the chamber wall, in most instances, continuously changes form. All we can do in this instance is to make up templates of the required shape at just a few positions around the chamber and then, by eye, blend these shapes into one another to produce a continuously and smoothly changing form. When grinding

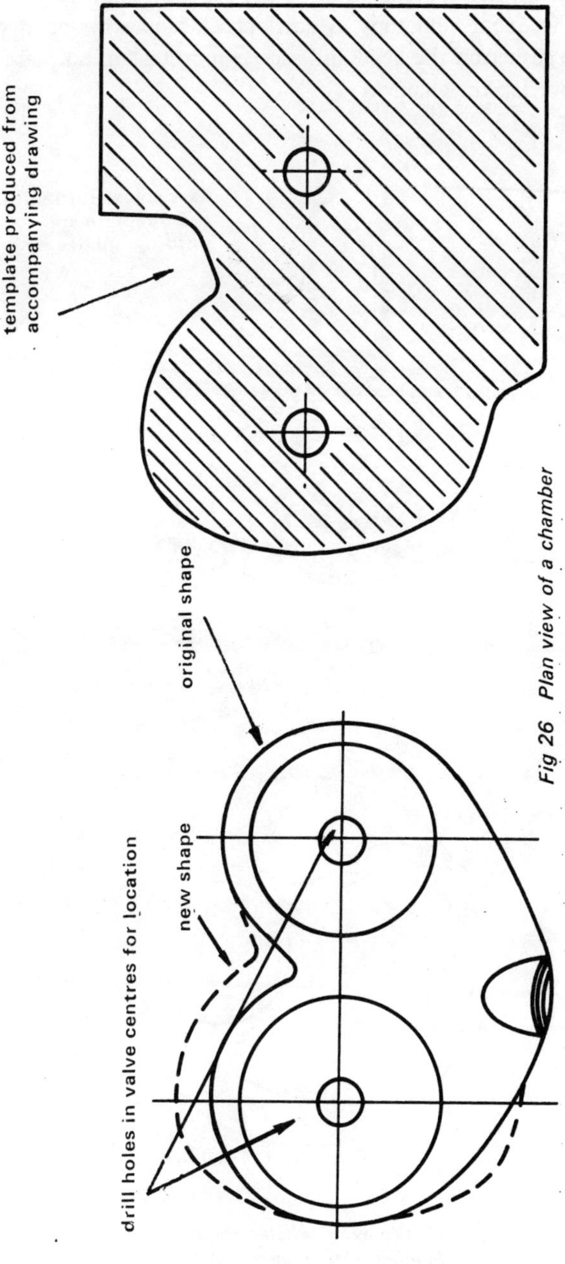

Fig 26 Plan view of a chamber

chambers, we must be sure not to undercut the gasket area (Fig. 28). If this area is infringed upon then you can expect to blow gaskets with regular and frequent monotony. A quick check can always be made by locating the gasket on the head and ascertaining that in no place does the gasket overlap the chamber.

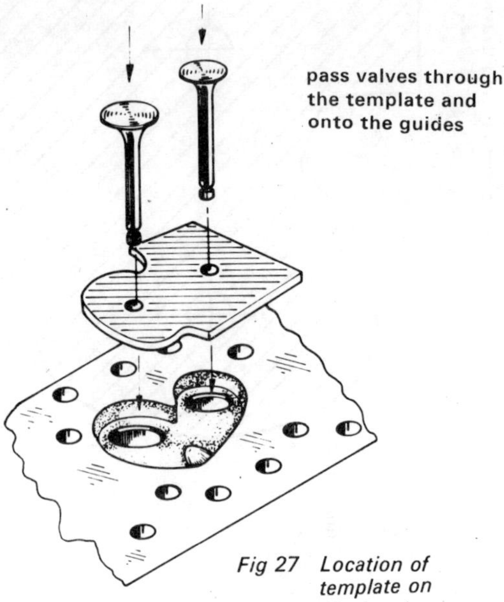

Fig 27 Location of template on the head

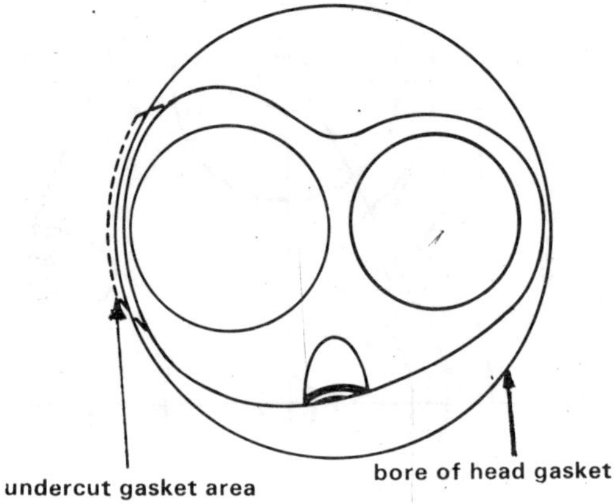

Fig 28 What to avoid when grinding chambers–an undercut gasket area

After the head has been skimmed, it is essential that all sharp edges are chamfered off. If this is not done, the edges can glow red hot and produce pre-ignition. To produce a smooth accurate chamber requires a little artistic talent and a lot of patience. Once the reader has got his eye in, chambers can be produced that need little or no balancing to bring them within the limit of 0.2cc as far as uniformity of volume is concerned.

PART TWO

chapter seven

Port Design

When tuning the average production engine, it is logical to set about the task in an orderly manner. The first aspect to consider is what modifications will give the greatest power increase or at least have the most significant effect on power when combined with other modifications. Since this chapter is on ports and port design, we will consider which of the two ports, inlet or exhaust, is the most beneficial to perfect first. Before that question is answered, we must consider the basic function of each port. The job of the inlet port is to pass air from the mouth of the carburetter or injector, which ever the case may be, into the cylinders of the engine. To be an efficient port it must do this with the least amount of pressure drop between the carburetter and cylinder, assuming a constant gas speed at this stage. This means that the higher the inlet pressure or the nearer it is to the external air pressure, the more power we obtain.

The prime function of an exhaust port is to pass the gases out of the engine and, assuming basic conditions again, this is done most efficiently when the least back pressure is created or presented to the exhaust.

Having established the basic functions of each port, we can now get down to establishing some criteria on which to base any mods. Using as a basis an ordinary production engine, we find that if we increase the inlet pressure by 1lb per sq in, ie simulate an increase in port efficiency, we get between 20 and 25 times the increase in power than that obtained by simulating an increase in exhaust efficiency by reducing the back pressure by 1lb per sq in. The obvious conclusion to draw from this is that the greatest gains in power are to be had by improving the inlet port or inlet system as a whole. In practice we find that the efficiency of an inlet system varies from engine to engine. Take, for instance, the inlet system of a twin cam Lotus-Ford engine. This has ports which have no abrupt curves, feed into a hemispherical chamber and each cylinder has its own carburetter barrel. All this goes to make up an efficient inlet tract. If the standard valves and carbs are retained, then a really good port job on the inlet ports only, results in a power gain of about 5%. In this particular engine, we find that there are greater inefficiencies in the exhaust system, mostly caused by the silencer. If the silencer is removed, we find a very similar power gain to that produced by a modified inlet port.

With a more mundane engine such as a series 'A' BL engine as fitted to Minis, A40s etc., increased induction efficiency whilst still retaining standard valves and carbs can result in a power gain of 30% whilst extensive work on the exhaust side will give only 10–11% power increase.

Since most production engines are not as sophisticated as the Lotus-Ford twin cam unit, we can expect by far the best results by concentrating a limited tuning programme on increased induction efficiency. If tuning is being done to achieve the greatest output possible, then obviously every aspect of increasing the power must be looked at.

Since the inlet port is the more important, we will deal with that one first. Before we get too detailed on this, it should be pointed out that other factors such as chamber shape, valve profile and the type of valve seat used have a distinct influence on the design of the actual port so not only will the port be dealt with but also these other relevant factors.

One of the first things the unenlightened head modifier does is to enlarge the inlet ports without thought as to why they need to be larger, if indeed they do need to be larger. Granted, a large diameter pipe will pass more air than a small pipe, but an inlet port is not as simple as a pipe. It has a reciprocating valve at one end of it and this tends to invalidate much, when a comparison is made between the two.

To determine the size of port required, we need a reference point to which we can relate it. In this instance the ratio of valve size to port size has a distinct effect on mass air flow. Thus if we know the valve size we are going to use, the optimum port diameter can be calculated. Fig. 29 is a graph which shows the discharge coefficient (C/D) of various ratios of port to valve head diameters. The diagrammatic layout shows the shape of the valve, port and chamber that produced the curves shown in Fig. 29. As can be seen, the best gas flow was obtained when the port size was made 0.8 times the diameter of one valve. This immediately dispels the theory that the bigger the ports are the better they are. Before we can apply the optimum port size shown by Fig. 29 to an actual port on an engine, we must consider that any variations of the test conditions will invalidate the results shown in Fig. 29.

You will notice in this figure that the valve opens on to a plain

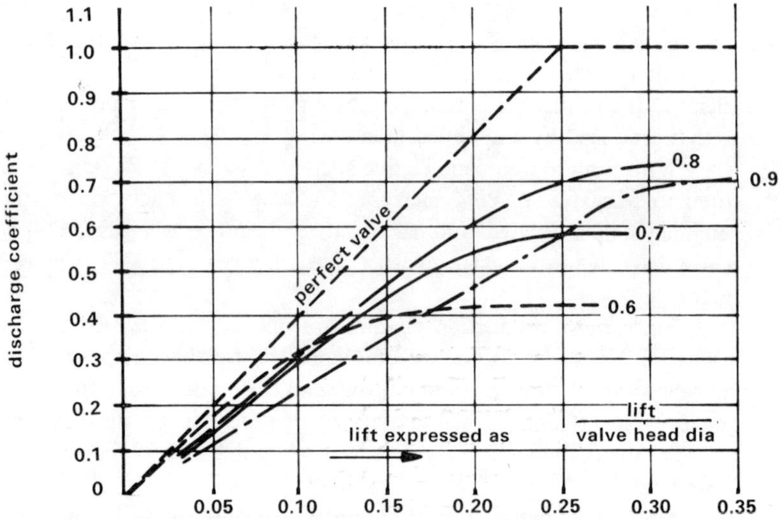

surface whereas in a conventional pushrod engine we have the chamber or cylinder bore wall in close proximity to part of the circumference of the valve. This causes shrouding and prevents a full flow of gas around the entire circumference of the valve. Fig. 30 shows how shrouding is caused by either the chamber wall or bore wall. The shrouding caused by a chamber wall need not bother us unduly, since this is one of the things we can reduce to a minimum. The shrouding caused by the bore of the engine cannot, in most cases be changed, especially if the bores are close together. Assuming the bores cannot be materially altered, then the shrouding so caused represents the lower limit of shrouding. When modifying a head, we must attempt to get the shrouding factor of the chamber equal to that caused by the bore. The shrouding factor, it should be pointed out, is a percentage measurement of the reduction in efficiency caused by the close proximity of the chamber or bore wall. An example here may serve to clarify matters. Take, for instance, a crossflow Ford head—since it has no chambers in the head, there cannot be any shrouding caused at this point. If we test the quantity of air that the head will flow while it is on the block, we find that it is 16.5% less than when it is flow tested by itself. The shrouding factor of this head then is 16.5%.

It may seem as if the subject has changed from ports to chambers. We are in fact still dealing with the topic of ports, but it is difficult to deal with one without making some reference to the other. As has been pointed out previously, the curves given in Fig. 29 are for a completely unshrouded valve. Since, in a conventional pushrod engine this unshrouded condition does not exist, we must make due allowances

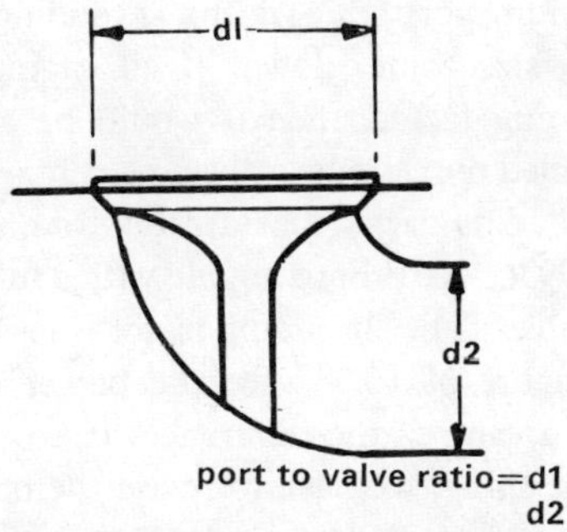

Fig 29 Determination of optimum port to valve ratio

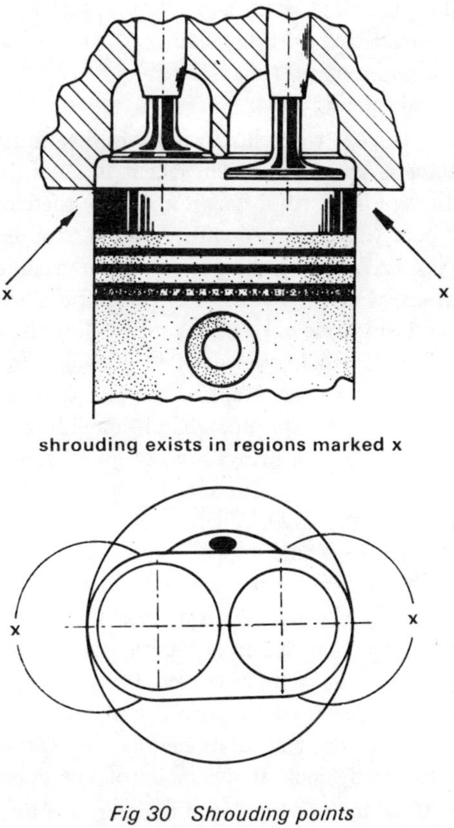

Fig 30 Shrouding points

to arrive at the optimum port size. As the shrouding factor goes up, so the optimum port size comes down. If all engines had the same lower limiting shrouding factor, then it would be simple to reduce our optimum unshrouded port size by a given percentage. Unfortunately the lower limit of shrouding varies considerably from engine to engine. Take, for instance, a 997cc 105E Ford engine with a fully modified head using standard size valves. The shrouding factor is less than 5%. When compared with the figure of 16.5% for the bigger crossflow engine mentioned earlier, you can see the differences can be quite large. All this boils down to one thing, we have to revise the optimum port size when the chamber is unshrouded to get the optimum port size when chamber shrouding exists. The size correction is quite simple, we make a percentage reduction on the optimum port size numerically equal to the square root of the shrouding factor times 1.4. For example, if we have a shrouding factor of 16% we reduce the port diameter from the optimum given in Fig. 29 by $4\% \times 1.4 = 5.6\%$ to put it into more mathematical terms, $CPD = OPD \times (1 - [1.4\sqrt{SF}])$

where CPD = corrected port diameter

OPD = optimum port diameter

SF = shrouding factor

The only trouble with this formula is that it assumes that the shrouding factor is known, when in fact it is quite difficult to put a figure to. Although primarily brought about by the close proximity of chamber and bore walls the shrouding factor is to an extent altered by the camshaft profile. The shrouding factor can be established reasonably accurately, either by a lengthy series of calculations which are strictly speaking beyond the scope of this book, or by flow testing under cyclic conditions. Since this book gives detailed drawings on how to modify heads of most of the popular cars, knowing the shrouding factor is not of great importance. On the other hand, should the reader wish to modify a head for which drawings are not given, then it is fairly safe to assume that the limiting shrouding factor is between 15 and 20% for an average conventional OHV engine.

So far the reader would appear to be safe in assuming that with a flat roof chamber, a circular port with one port per valve the optimum gas flow would be achieved by using the formula 0.8 times valve diameter times correction due to shrouding, and so you would, on a steady state gas flow machine. However, few of us are likely to drive around with a cylinder head fitted to a gas flow machine, the usual thing is to fit it onto the rest of an engine. To expand on this it will be necessary to briefly look at the mode of operation of a gas flow machine. The usual method employed to determine the flow characteristics of a head is to draw or blow air into or out of the relevant port. If the test is being done on an actual head, then a cylinder is placed under the head to simulate the cylinder bore of the engine (Fig. 31). The gas flow is then tested with the inlet or exhaust valve, whichever the case may be, opened by various increments up to the full lift value Under these conditions the air flow is said to be in a steady state condition since it is neither accelerating nor decelerating and these conditions do not exist when the head is on the engine.

Under actual conditions, the air flow in the inlet port is pulsating. It accelerates as the valve opens and reaches its maximum speed just after the piston reaches its maximum speed down the bore. As the piston reaches the bottom of its stroke, the speed of the air flow drops. When the piston passes bottom dead centre, and starts its journey up the bore, the air flow into the cylinder, especially at high revs, does not cease. That is, of course, assuming that the inlet valve is still open. The reason for this is that at high revs, the piston travels down the bore so fast that it leaves the air in the port behind. This causes a partial vacuum in the cylinder which can continue to be filled whilst the piston is starting to come up the bore. Apart from this, air has inertia and is reluctant to stop. Because of this reluctance to stop, the air can go on piling into the cylinder even though the piston is on its upward journey.

What has all this to do with our inlet port design? Simple. We wish to design the inlet port such that the air it carries contains the maximum energy, and that we use this energy in the best possible way to fill the cylinder. To this end we find that if the valve throat is enlarged slightly, the air flow is increased. The optimum throat diameter would appear to be in the order of 12–14% greater than the optimum port diameter. For example, if the optimum port diameter proved to be 78% of the valve diameter, the throat diameter should be 78 + 9.5% of valve diameter = 87.5% of valve diameter. The enlarged valve throat should extend about 1in from the valve seat and then blend smoothly into the rest of the port. It should be pointed out that it is also essential for the valve throat to blend smoothly into the valve seat. This point is critical as a measurable amount of power can be lost if at this point the valve seat forms a sharp edge with the valve throat.

At this point we can build up a picture of what our port should look like, bearing in mind the conditions that have been laid down. To go over these conditions again, we have one port per valve. The port is circular, the valve seat is a conventional 45° seat and the chamber roof is flat. In Fig. 32 we have a port which, for the sake of clarity, has been straightened out. This shows the valve seat, the enlarged valve

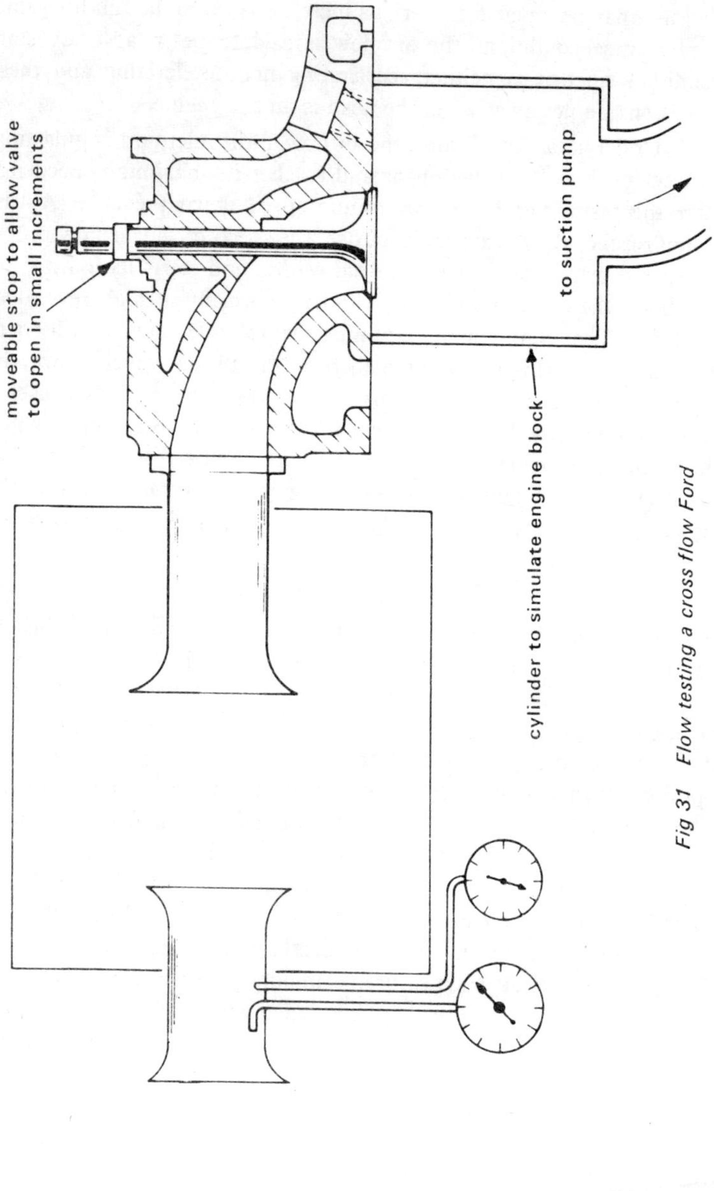

Fig 31 Flow testing a cross flow Ford

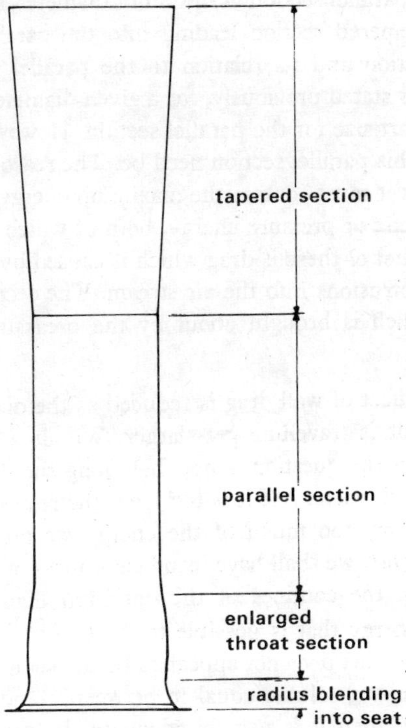

Fig 32 A straightened inlet port showing the various changes in form

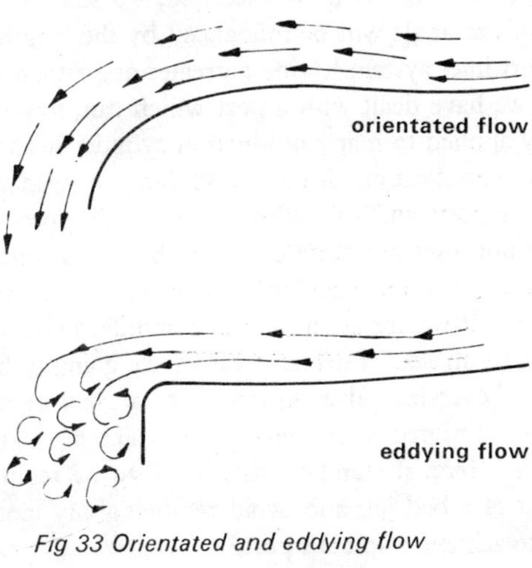

Fig 33 Orientated and eddying flow

throat, the parallel section at optimum diameter and one more thing—a shallow tapered section leading into the parallel section. It is the tapered section and its relation to the parallel section we will now consider. As stated previously, for a given diameter of valve there is an optimum port size for the parallel section. However, there is a limit to the length this parallel section need be. The reason for this stems from the fact that it must contain the maximum energy possible; two things destroy kinetic or pressure energy, both of which are interrelated.

The first of these is drag which is caused by the walls of the port and any protrusions into the air stream. The second one is caused by eddying which is brought about by the break-up of orientated flow (Fig. 33).

The effect of wall drag is reduced as the diameter of the port in which the air is travelling gets larger. We already have an optimum port size, but the question is now how long should the port remain at its optimum diameter? If it is too long, the restriction caused by drag will drain away too much of the energy we are trying to preserve. If it is too short, we shall have insufficient mass of air travelling at high speed within the confines of the optimum diameter to contain the maximum energy that is possible to store. The length of the parallel section of the port does not appear to be too critical. If it is made such that it contains a volume equal to between $\frac{1}{8}$ and $\frac{1}{5}$ of the volume of one cylinder, then it is usually adequate. In practice this works out quite conveniently, for with many heads we find that the parallel section is contained in the head whilst the tapered section is within the carburetter or injector manifold. The included angle of the tapered section should be within 4°–8°. If 8° is exceeded, we start to lose out. The final choice of the angle will be influenced by the length of the intake system, a short inlet system having a greater angle than a long one.

So far we have dealt with a port which can, to a certain extent, be practically applied to many production cylinder heads. To produce a port to the specification discussed so far, we need a head which initially has the ports and valve throats too small. Ports which are too small for our purposes are quite common, but heads having undersize valve throats when using standard size valves are more of a rarity. This means that if we are going to retain standard size valves, a compromise must be struck. If oversize valves are going to be fitted, then the problem of oversize valve throats is, to a great degree, removed. Some heads do, unfortunately, have ports which are too large for the standard valve. Here, if standard valves are to be retained, we must make the best of a bad job and avoid removing any more metal than is necessary to achieve a smooth port.

On occasions, head designers use square ports, usually because there is insufficient room to accommodate round ports of adequate area. When confronted with a square port where one port feeds one valve, we should endeavour to achieve a cross-sectional area 10% greater than the equivalent round port, the reason being that a square port restricts air flow in comparison with a round port, due to increased wall drag and turbulence. The increase in area over the optimum size round port does, to a small extent, offset this.

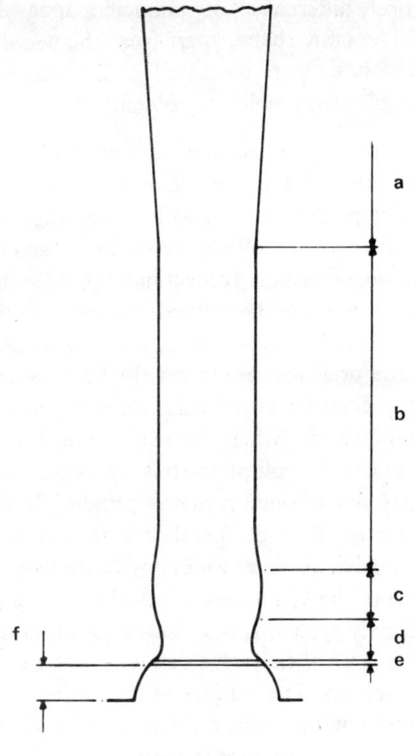

a tapered section
b parallel section
c enlarged throat section
d venturi section
e valve seat
f chamber wall

Fig 34 Make-up of venturi port, straightened out for clarity

Although the type of port so far dealt with represents a fair improvement over many standard ports, it is by no means the best that can be produced with present knowledge. The venturi port, which we are about to discuss, does show a significant increase in air flow over the more basic type of port discussed so far. One of the most important parts of an induction system as far as breathing goes, is the area just before and just after the valve seat. It is by subtle changes in this region that the venturi port gains in efficiency when compared with a more basic port. Fig. 34 shows the basic make-up of a venturi port with its form exaggerated for the sake of clarity. With this type of port we have an entirely different set of parameters upon which our optimum port is based. The basic shape, apart from the venturi section, will be as for the more basic port discussed earlier but we have a greater number of variables with which to contend.

Starting at the tapered section, we find that this will be the same as the previous case, that is, an included angle between 4° and 8°. The length of the parallel section will also be the same at $\frac{1}{8}$ and $\frac{1}{5}$ of the volume of one cylinder. When we come to the optimum diameter we find the first sign of change. Inspection of the relevant graph (Fig. 35) will reveal that if we use the curved chamber roof together with a venturi port, the optimum diameter of the parallel section needs to be increased over the previous case to get the best results. With a curved chamber roof the diameter of the parallel section of the port should be 0.84 times valve diameter when the valve is unshrouded. To get the port size for a practical application, this figure should be corrected to take into account any shrouding that is present. If a curved chamber roof isn't being used, then the parallel section of the port should be about 0.82 times valve diameter times any correction due to shrouding.

Moving on to the valve throat, or to be more precise, the part of the port immediately preceding the venturi, we find the diameter should increase to about 0.97 of the valve diameter and then smoothly blend into the venturi section. The venturi itself is quite a complex affair and after experimenting with various combinations of diameters, angles and radii, it was found that the profile shown in Fig. 36A gave the best results. The valve seat life, however, was short, acceptable for a racing engine, but not so for a road engine. The wear suffered by the seat did not impair the sealing but did have the effect of deforming the radius, thus destroying the flow properties of the seat. To achieve a longer seat life the profile shown in Fig. 36B can be used. It is only marginally inferior to type A but does not suffer such rapid seat deformation. On the debit side, seat B is much more difficult for the amateur to produce than seat A.

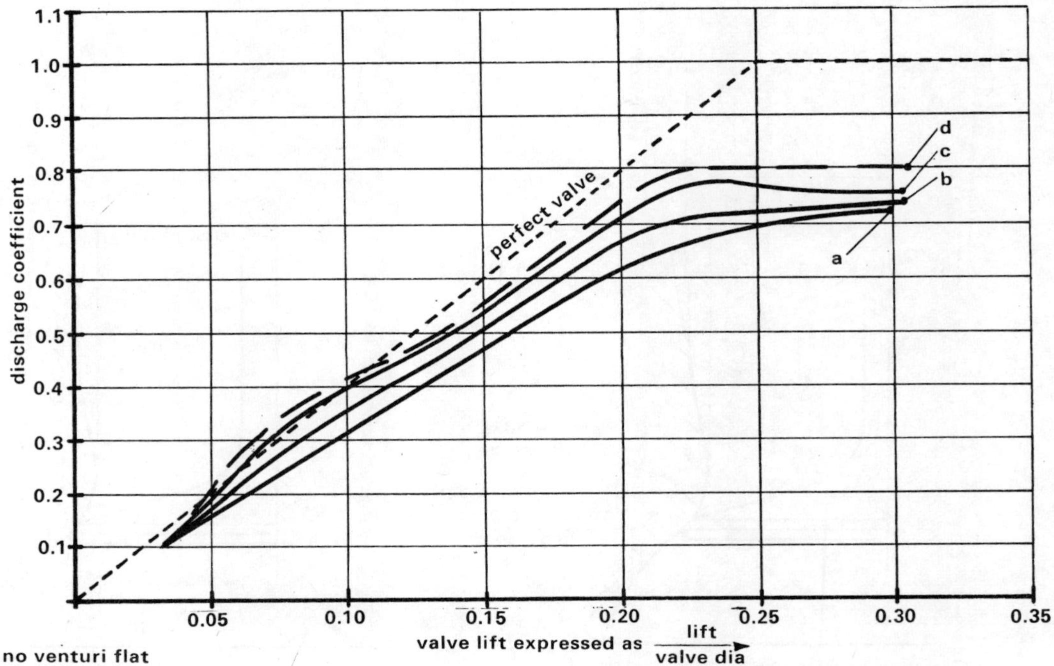

a a conventional 45° seat no venturi flat roof 0.8 port/valve ratio

b venturi port with flat roof 0.82 port/valve ratio

c venturi port with curved chamber roof 0.8 port/valve ratio

d venturi port with curved chamber roof 0.84 port/valve ratio

Fig 35 Optimum port to valve ratio for venturi port with curved chamber roof

Fig 36A Valve seat profile—racing engine

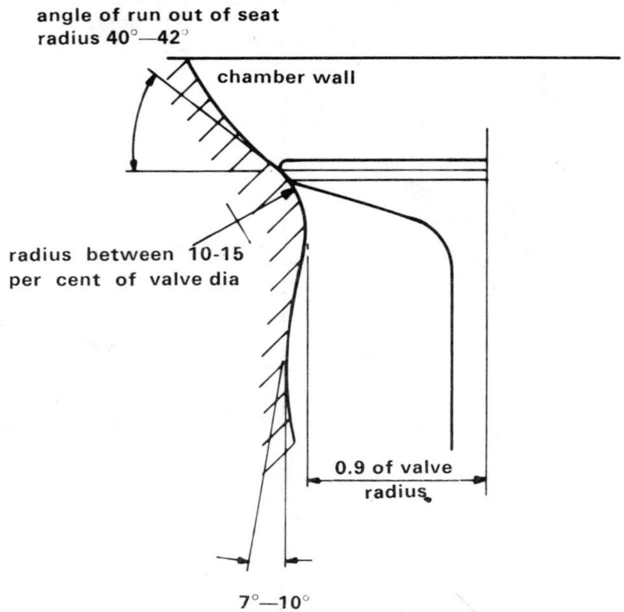

Fig 36B Valve seat profile—road engine

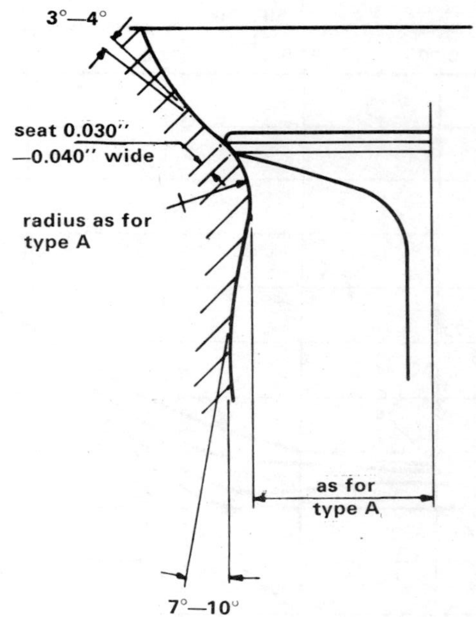

The curved type chamber roof as shown in Fig. 36 has quite a distinct effect on gas flow and for this reason it will be dealt with as a part of the port. If this curved section joins the seat at the right angle and is of the correct profile, it can play an important role inasmuch as it reduces turbulent and eddying flow in the vicinity of the seat. Also, the air is not so violently decelerated upon entry into the cylinder as it is with a flat chamber roof. All this goes to ensure better cylinder filling which inevitably results in more power. The correct curve for the chamber wall is given by a mathematical equation. In practice we find that this curve is not too critical. If it is produced as in Fig. 36, it will function more or less as intended. Apart from its very noticeable effect on gas flow, the use of this curved chamber wall brings about other advantages as far as the comubstion process is concerned. These, however, will be dealt with in a later chapter concerning combustion chambers only.

Up to now we have looked at ports which are more or less ideal inasmuch as they have only one valve per port. As many people well know, there are numerous cars on the market which have siamesed ports, that is one port feeding two valves. This means that much of the subject matter concerning port sizes no longer applies, since we are now operating under a different set of conditions. To start with, a single port has to contend with the gas flow requirements of two cylinders instead of one and for this reason one would expect it to be larger than when one valve per port is used. This, in fact, does prove to be the case, but it is not, as one would at first assume, twice the size of a port feeding one valve. The graph (Fig. 37) shows that the mass air flow under cyclic conditions is greatest when the port diameter of the siamesed section is between 1 and 1.1 times the area of one inlet valve. At this point it should be made clear that it is all too easy to draw an incorrect conclusion from this piece of information just given.

If we have an inlet port which is already the optimum size for a given valve and the port cannot, for physical reasons, be enlarged any further, it does not mean that the air flow cannot be increased by putting in larger valves. The reason for this is quite simple. The efficiency of the siamesed section of the port is far higher than the valve and valve seat as far as air flow is concerned. When the valve size is increased, we have a greater area through which to draw air but because the port size cannot be increased, the efficiency of the complete system drops off. If we keep on increasing the valve size in increments, we find that the time comes when efficiency is falling off faster than the valve area is increasing. When this point is reached, we can no longer get any increase in air flow. The point at which it appears to be impractical to increase

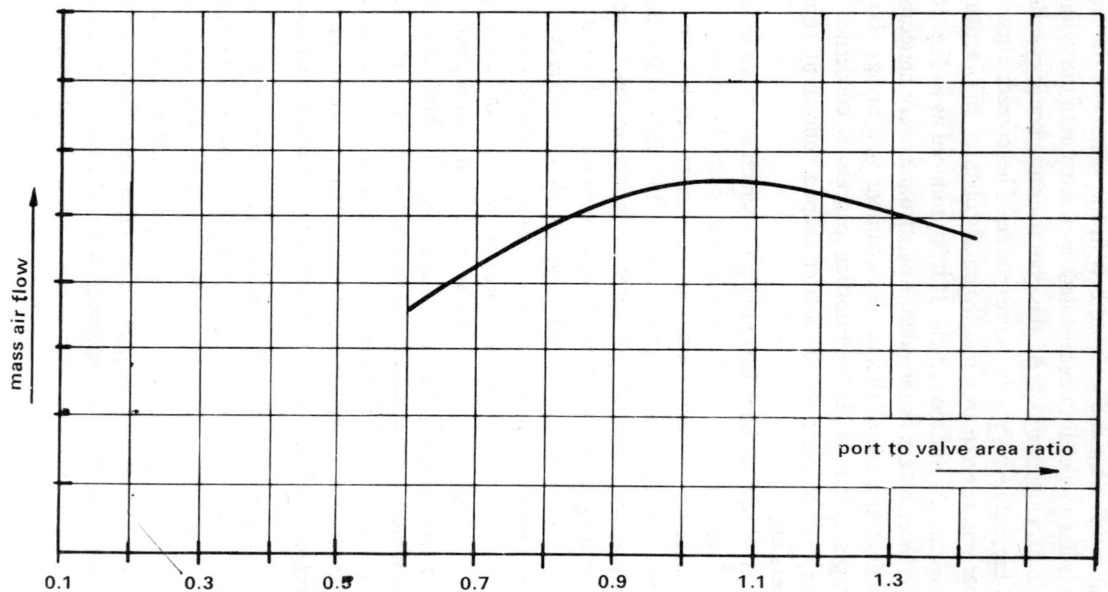

Fig 37 Determination of optimum port to valve ratio with siamese ports

the inlet valve size any further occurs when the area of one valve is 50% greater than the area of the siamesed section of the port. As far as venturi valve throats, radius seats and curved chamber walls are concerned, these will apply more or less as before.

Another factor vital to good gas flow is the shape of the inlet valve. In days past, it was thought that a well-tuliped valve gave the best gas flow by virtue of its streamline shape. This, however, was a fallacy. It is, in fact, possible to show by geometric means that if the underhead radius exceeds 0.25D (where D is the valve dia.) the part of the underhead radius that is level with the valve seat is causing a reduction in valve throat area (see Fig. 38). If that isn't clear, then the best way to visualise it is to assume that the underhead radius is very large, say about equal to the diameter of the valve.

Fig 38 Valve profile

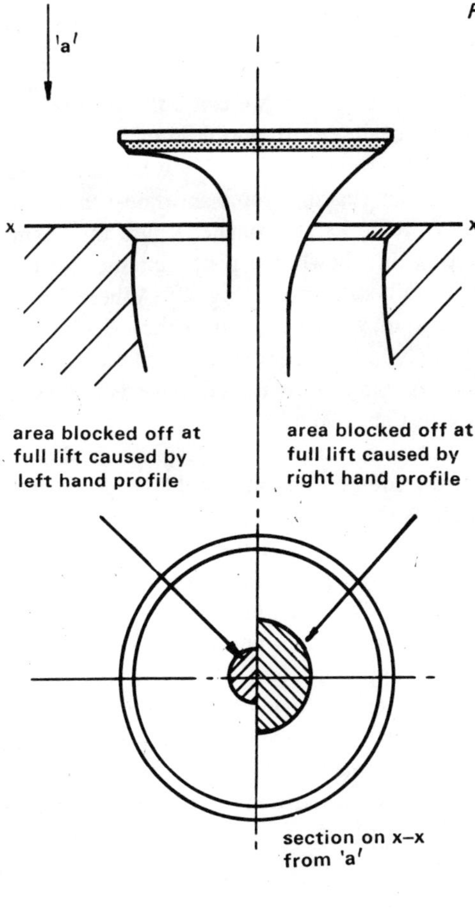

When the valve is at full lift, an appreciable amount of the tulip section is still within the confines of the valve throat. Since the tulip diameter, which is level with the valve seat at full lift is of larger diameter than the valve stem, it is, in effect, partially blocking the hole through which the air is expected to pass. To avoid any such restriction, the underhead radius must not exceed 0.25 times valve diameter. Apart from attention to the tulip part of the valve, we can, by careful attention to profile, achieve a secondary venturi action at the valve seat which increases the discharge cofficient considerably at low valve lifts. This is of considerable importance, since it achieves the same effect as using a cam with a faster opening and closing motion with a conventional seat and port. The advantage of this secondary venturi effect is that we achieve the gains of the camshaft without any extra stresses imparted to any of the relevant components. Aside from any considerations of gas flow, the valve must be strong enough to withstand the loads expected of it, yet be as light as possible to keep the valve spring pressure to a minimum.

Using the information given so far, we can build up the profile of a valve for use with an ideal venturi port leading into a curved chamber roof, giving the form shown in Fig. 39. When we come to less ideal situations such as flat chamber roofs and non-venturi ports etc, the theoretically ideal profile has to change. When the chamber roof is completely flat, as with a bowl in a piston engine, we find a flatter underhead profile gives better results (Fig. 40). Where the combustion chamber lies in the head, we can easily have the situation where the circumference of the valve is surrounded by a shrouded area, a curved roof and a flat roof. A compromise therefore needs to be struck between the valves shown in Figs. 39 and 40.

Having considered various lengths, diameters, profiles and shapes of ports and valves, some attention should be given to the routing of the port. Without a shadow of doubt, the straighter the port is, the better the gas flow will be. Since it is always necessary to have a bend in a port, it is a good idea to make its radius of curvature as large as possible. Strictly speaking, it is a good idea to make sure that the curvature of the inner and outer sides of the port have a common centre. If this is not done, we can, by applying a larger radius to the inner side of a bend, induce an earlier onset of eddying flow (see Fig. 41). If, on the other hand, the radius of the inner edge is very tight, then it can pay to enlarge it, even thought it may not then have a common centre with the outer edge of the port. This is just one more of those situations where compromise is the password.

Apart from the general shape of a port, the surface finish can

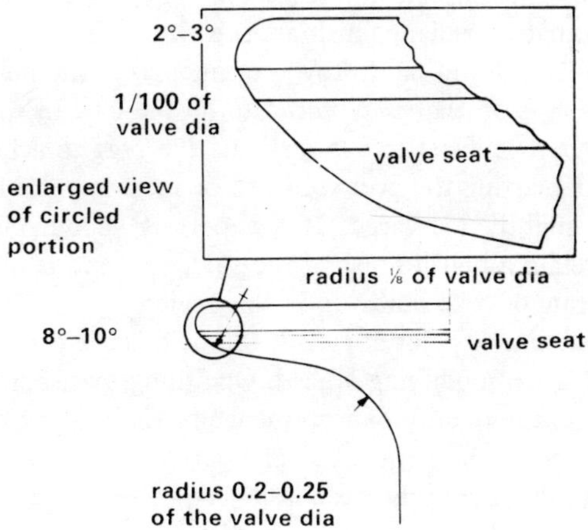

Fig 39 Valve profile for use with venturi ports and curved chamber roof

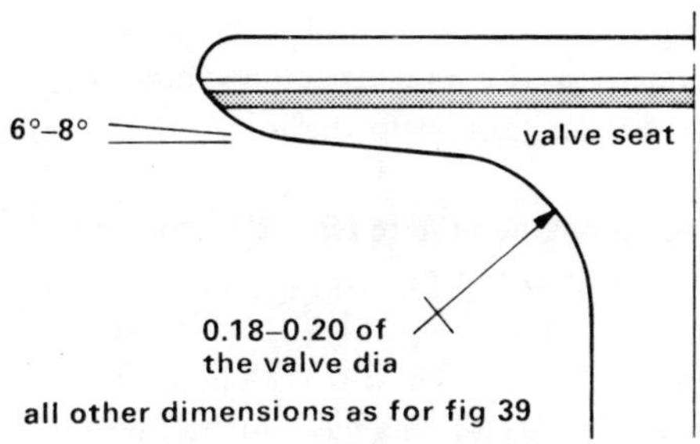

Fig 40 Valve profile for use with flat chamber roof

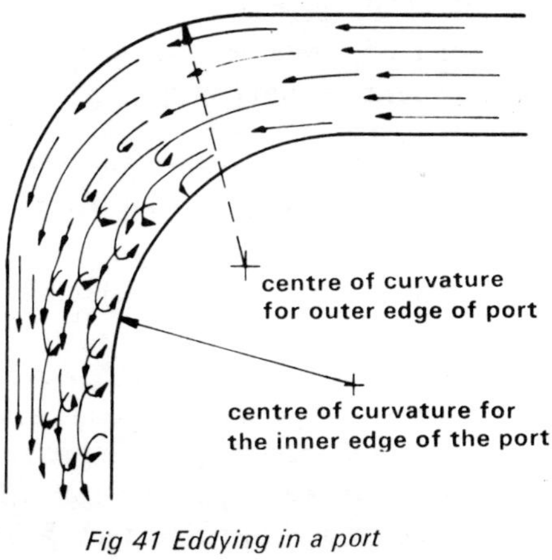

Fig 41 Eddying in a port

bring to bear a measurable influence on the gas flow. The popular conception is that a brilliant finish akin to chrome plating is the best. This is not so. It can, in fact, prove detrimental to the power output. A smooth finish is all that is required to do the job. In some cases it can prove to be beneficial to vapour-blast the port to achieve a matt finish. Vapour-blasting the port seems to be most applicable when the inlet port is slightly too large. If the port is vapour-blasted, it is advisable to retain a polish finish at the valve seat and its near vicinity to retard the rate of coke build-up in this region.

From the tuner's point of view it will be necessary to accept many compromises, when modifying a head. One thing that should never be compromised is the quality of workmanship. As we can see, it needs only small changes in the form of the inlet port to give less than optimum results as far as the part of the head is concerned. Because of this, every effort should be made to produce as good a job as possible for it is of little use having the knowledge if it isn't put into practice.

Exhaust ports

The function of the exhaust port is to convey the burnt gases from the cylinder to the outside atmosphere with the minimum hindrance. In a similar fashion to the inlet valve, we find that careful consideration of port shape and size can beneficially influence the power output of an engine. There is, within the cylinder of an engine, quite a considerable pressure prevailing when the exhaust valve opens. Because of this, the initial flow in the exhaust port can be likened to a slug of gas leaving the cylinder and travelling down the port at high speed. The kinetic energy of this gas can be used to help extract the remaining exhaust from the cylinder, thus relieving the piston of some of the work, as well as assisting in scavenging the chamber when the piston reached TDC. If the port is too big, the speed of the gas will drop and the extraction effect will be reduced or even lost. Should the port be too small, the restriction to flow will be high, which again will lead to inefficient exhausting. The only solution is to strike a balance between the two extremes. This is not as simple as it at first may seem, for the final design of the exhaust port and the exhaust system as a whole is affected by the camshaft timing. On top of this, we can, as with the inlet valve, experience exhaust valve shrouding which can lead to further complications.

Let us first consider the effect of the camshaft timing. As the cam timing gets longer, ie the opening and closing points occur sooner and later in the cycle, so the pressure prevailing in the cylinder when the exhaust valve opens, increases. Also the longer cam timing implies that the engine is to operate at higher revs, thus the speed of opening of the valve with respect to time, is faster.

From these two points we can see that the initial slug of gas from the cylinder will occur with a more well defined front and will be travelling at a higher speed. This means that as cam timing is increased, so the amount of exhaust handled in a given time is increased during the critical exhaust blow-by period before the piston forcibly expels the gases. As one would expect, this dictates the need for a bigger bore exhaust port or exhaust pipe than is required for an engine using a cam with shorter timing. In practice, the exhaust port diameter is between 1.15 and 1.25 times the diameter of the exhaust valve, the lower figure being applicable to road cams while the higher one is intended for competition cams. A cam with semi-competition timing will, of course, fall somewhere between the two extremes. Obviously if the exhaust port is to become larger than the valve, it must taper outwards from the valve size or rather, to be more precise, from the inner valve seat diameter to its intended size. If the port tapers out too quickly, the flow will inevitably break away from the port wall. This will cause a loss of energy due to increased turbulence and a rise in back pressure. An ideal angle to taper the exhaust port to is between 2° or 3° inclusive, but in many cases we find that the port does not reach its intended size before the exhaust manifold is reached. This need not be a disadvantage if the exhaust manifold is made to suit. On the other hand if the exhaust port has to be matched to a proprietary manifold of the correct bore size, the taper angle of the port may well have to be increased. The approximate maximum angle of taper before the flow breaks away from the port wall is between 7° and 8°, and this should be regarded as the upper limit for exhaust port taper.

A reduction in the required port size is brought about by any shrouding that may exist at the exhaust valve. The reduction in diameter should be in proportion to the square root of the shrouding factor, ie if the shrouding factor is 9% the port diameter reduction will be $\sqrt{9} = 3\%$.

When a head has siamesed exhaust ports, a further size increase is required in the port at the siamesed section. The area increase over that of a single valve per port is somewhere between 25% and 40% depending again on the type of camshaft used and, to a lesser degree, on the exhaust manifold design. If the length of the exhaust pipe from a siamesed port is relatively long before it joins up with others, the optimum exhaust port area comes down slightly. If the length of the exhaust pipe to its first join is progressively shortened, then the optimum port area goes up slightly.

Curves in an exhaust port should be dealt with in the same

fashion as the inlet, ie the radius of curvature of the outside and inside should have the same centre. Only when the inside curve is very tight should one attempt to enlarge it. Even so, this must be done with caution so as not to overdo things.

chapter eight

Combustion Chamber Design

The brief study of ports in the previous chapter will have given the reader a fair indication of how to fill and empty the cylinders. Having filled the cylinders as full as possible, it is essential that the mixture of fuel and air is burnt as effectively as possible, and that the heat energy produced is converted efficiently into mechanical energy. If this is not done, then no matter how well the engine breathes, it still will not produce its full power potential.

When the breathing ability of an engine is referred to, one often thinks of streamlined ports, a multiplicity of carbs, sweeping exhaust manifolds and so on, but let us not lose sight of the fact that the combustion chamber shape also has a profound effect on breathing. In the previous chapter on port design, you will note that references have been made to valve shrouding, which is a situation that exists when a chamber wall or bore wall is in close proximity to the edge of the valve head. It was also stated earlier in the same chapter that the lower limit of shrouding is set by the proximity of the bore wall to the valve head and not the chamber wall to the valve head. The primary reason for this is that if we have excessive shrouding caused by the chamber wall, we can cut away the part of the chamber wall which is causing shrouding (Fig. 42). The removal of the offending part of the chamber wall can continue until the shrouding factor of the chamber is reduced to that caused by the bore wall. At this point we can make no further reduction in shrouding assuming that the bores cannot be materially altered.

Providing things have been understood so far, we can go one step further. What happens to our shrouding factor if the valve size is altered? The answer to this is straightforward enough. If the valve size goes up, the shrouding factor goes up. The net result of this is that although we may have increased our valve area by, say, 40%, the shrouding factor may have increased by 10% so that we do not gain the full benefit of the 40% increase in valve area. Remember the shrouding factor that is being referred to is the lower limiting one, that is, the one caused by the bore. The reason this has been stressed

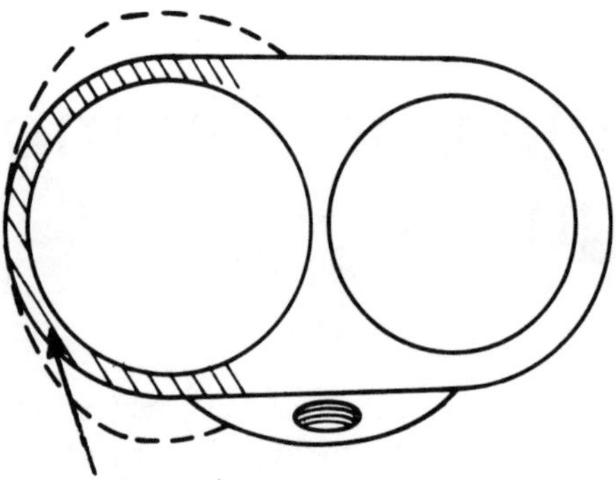

Fig 42 Shrouding caused by chamber wall

is very important. If a larger valve is fitted to a chamber which is remaining standard, the shrouding factor caused by the chamber can go up faster than the valve area! If this is the case, then the larger valve will pass less air than the standard one, so this demon mod has, in fact, reduced the potential power output of the engine. From this we can see that one simple rule can be applied and that is never fit larger valves into a head unless the chamber shape is modified to keep the shrouding factor at its lower limit. Even if this rule is applied, there are some instances where the shrouding factor caused by the bore imposes an upper limit before all the room for larger valves is used up, but these cases are few and far between. Graph number 1 in Fig. 43 shows how quickly the lower limiting shrouding factor can rise as the valve size is increased and this rate of increase is typical of a conventional pushrod engine where all valves are in line.

Amongst the more modern designs of pushrod engines there is a trend to incline the valves towards the bore centre line (Fig. 44). This has the effect of reducing the shrouding factor, thus allowing the head to breathe better. This method of inclining the valves is, strictly speaking, splitting the difference between in line valves in a bath tub chamber and the steeply inclined valves of a hemispherical chamber layout which, incidently, suffers from the minimum of shrouding. A hemispherical chamber can and does in some instances go to zero shrouding.

Just to show how important valve shrouding is, let's look at an example. Many people, if confronted with two alternative ways of getting more power from, say, a Mini engine, would choose the fitting

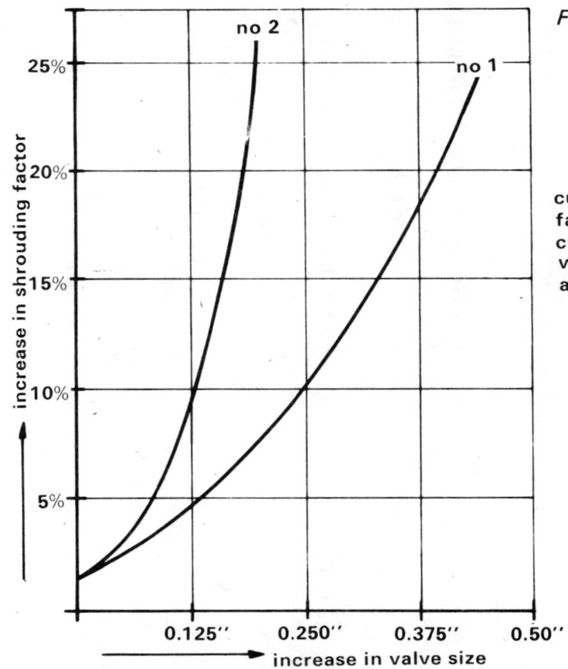

Fig 43 Effect of increase in valve size on shrouding factor

curve no 1 lower limiting shrouding factor ie that caused by the bore
curve no 2 shrouding factor curve when cnamber is not reshaped to accept larger valves

of a larger inlet valve rather than the modifying of the chamber shape to unshroud the valves. Let's consider in detail the effect of each of these modifications on the effective valve area.

A standard 848 Mini has an inlet valve of $1\frac{3}{32}$in dia. and with the standard chamber shape the shrouding factor is 36%. The standard valve size with a 36% shrouding factor thus only breathes as well as a 0.860in dia. valve with zero shrouding. If the valve size is increased to the size used in BLMC 1100 engines, that is $1\frac{5}{32}$in dia., the shrouding factor goes up to about 44% in a standard chamber and the valve breathes only as well as one which is about 0.840in dia. with zero shrouding. From this we can infer that fitting larger valves into a Mini head, whilst leaving the chamber shape standard, is an effective means of reducing the breathing potential and hence the power output of the engine. By cutting away the chamber around the inlet valve, we can reduce the shrouding factor of 36% with a standard valve down to the lower limiting shrouding factor caused by the bore, that is 16%. This means our standard valve now breathes as well as a zero shrouded valve of approximately 1.010in instead of one of 0.860in. By removing the shrouding, we have gained something like 25% in effective valve area.

This very rapid rate of increase in shrouding with increasing valve sizes in a standard chamber is typical not just of a Mini engine, but many other engines as well. Graph 2 in Fig. 43 shows the rate of increase in the shrouding factor on a Ford 105E head as the valve size increases.

The exhaust valve can, like the inlet valve, be shrouded, but since it is a common thing to use smaller exhaust valves than inlet valves, most standard production heads do not suffer such a great degree of exhaust shrouding as inlet shrouding. Any shrouding that does exist is, however, treated in a similar fashion to inlet shrouding, that is by cutting away the walls of the chamber.

So far it would appear that all we have to do to get a good chamber, is to grind out large quantities of metal from around the valves. Unfortunately this is not so. Before putting grinder to head, we must consider what, if any, adverse effects may be caused by grinding away large quantities of metal. The most obvious point is that the removal of metal from the chamber will lower the compression ratio. Skimming the head will raise the CR but if too much metal has been taken out of the chamber, we may find that the head will not stand skimming enough to raise it back to a reasonable level. Cutting back the chamber walls shows a diminishing rate of returns, that is, if the head is cut away around the valves in even stages, the first stage

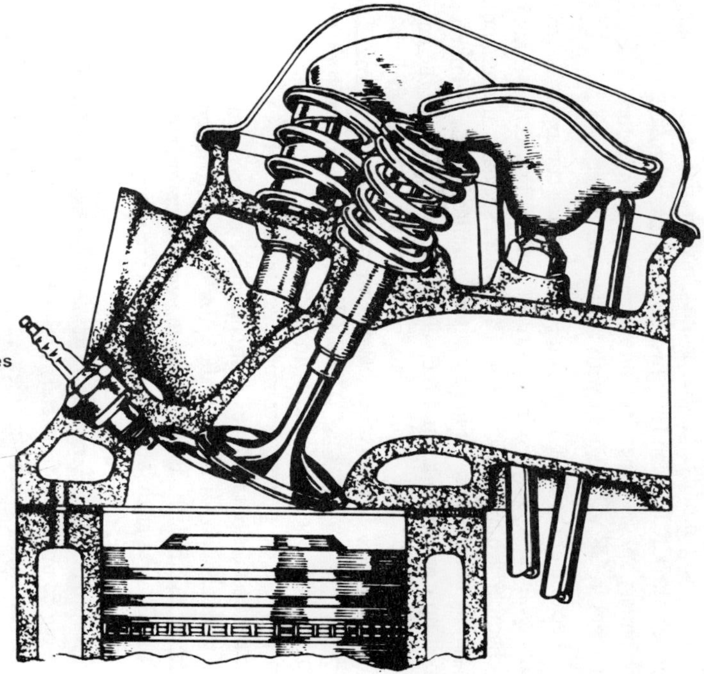

Fig 44 The Chevrolet Porcupine head

will show a greater return in power than the last stage. The same goes with the CR. Each rise in CR shows less gain than the one before. With any head modifying, the prime consideration is the breathing potential, the CR is secondary. Just how important the CR may be in its secondary role is dependent on numerous other factors. The effect of changes in CR will be dealt with later in the book.

Because of the diminishing rate of return for cutting away around the inlet valves, it is necessary to strike a balance between opening out the chamber around the valves and the CR, especially if the amount that can be skimmed off the head is severely limited. If the chambers are opened up excessively around the valves, we may find that the engine breathes well but loses some of its potential power through lack of a high enough CR. On the other hand if we relieve the area around the valves too little, we may get a high CR but due to lack of breathing, the piston will not have a full cylinder of air to compress and we will again lose out on power. A compromise, therefore, needs to be struck to get the best output from any particular head unless it can be skimmed more than enough. If we were designing a head from scratch, then it would be a simple matter to cater for good breathing whilst making sure the CR was at the required level. We are not, however, designing a head from scratch but modifying an existing one, so these compromises are inevitable; indeed the whole business of head modification is one big compromise from start to finish.

If the head being modified can only stand a small amount of skimming, we restrict the amount cut away around the valves to conserve the CR but it should be stressed that the bias towards the retention of CR is small. As has been stated before, the CR is of secondary importance and only assumes major importance in terms of a lower limit below which it cannot be allowed to fall. It is possible to calculate within reasonable limits of accuracy at what point the balance between CR and increased breathing by unshrouding the valves will give the best results in terms of power. An easier method than calculating is to raise the CR as high as it will go so long as it is still under the point at which detonation occurs, and then progressively unshroud the valves until the point at which the best results are seen. Obviously to find the best point it will be necessary to go too far with the unshrouding. This means two heads will be needed, one as a try-out which will be scrapped, since the shrouding will have been taken too far, and the second head, which is ground to the point which gave the best results that were achieved on the first head. If we have a head which will stand a large amount of skimming, then the problem of reaching the desired CR after unshrouding the valves fades into insignificance. All that need be done is to grind the chamber shape to give optimum breathing, at this point no other factors affecting power are being considered, and then skim it to give the required CR.

The breathing ability and the selection of suitable CR are not the only factors governing the power output of the engine. When valves are unshrouded we (apart from changing the CR, which in many cases is redeemable), alter the squish area (Fig. 45). Before it can be said whether this is desirable we must consider the function of the squish area of a combustion chamber.

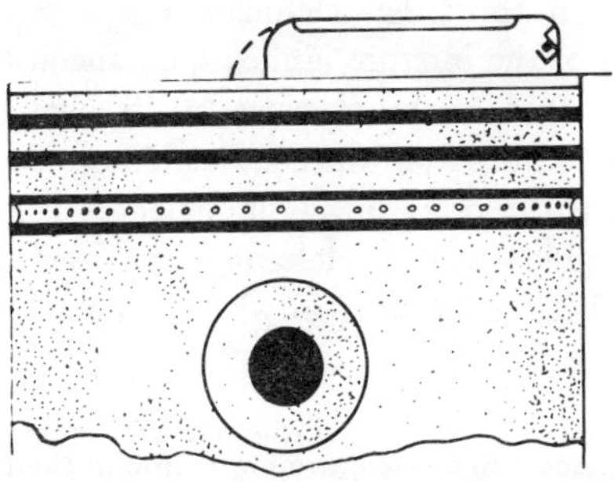

Fig 45 Reducing shrouding

Imagine a long glass tube with a bore of about $\frac{1}{2}$in-$\frac{3}{4}$in dia. and open at both ends. If this tube has a mixture of petrol and air in it, which is stationary, then we would find that if the mixture was ignited at one end of the tube, the flame front would travel down the tube at a rate barely equal to that of walking speed. Obviously if such conditions existed within the cylinders of an engine, the mixture would have hardly started to burn before the exhaust valve opened to rid the cylinder of its charge. From this we can see that it is of prime importance that the charge is burnt fast enough to be completely burnt by the time the exhaust valve opens.

Assuming that the fuel air/ratio supplied to the engine is right for maximum power, there are four factors which increase the rate of burning. These are:

1. Increasing the density of the charge,
2. Increasing the temperature of the charge,
3. Increasing the pressure of the charge at the time of ignition,

4. Producing turbulence or swirl within the charge.

When looked at from the point of view of an internal combustion engine, the first three points are to a large extent interrelated. For instance, if we increase the pressure of the charge by raising the CR we inevitably increase the temperature and density of the charge at the time of ignition.

Although raising the CR to increase all three of these factors significantly increases the rate of burning, it is invariably insufficient for complete or as near complete, burning as possible. Apart from this, there are some practical limitations which impose an upper limit on the use of these factors. For instance, if the temperature of the charge is too high we can, even if the cylinder pressure is more or less reasonable, get detonation occurring. Detonation is a phenomenon which occurs when part of the mixture ignites spontaneously throughout its volume (Fig. 46). It does this because (a) its own temperature and pressure rises due to the part of the combustion cycle which has already taken place, and (b) it picks up additional heat from the advancing flame front. An engine which is detonating runs very rough and because of the high local heating in the area where detonation is occurring, it quickly destroys itself.

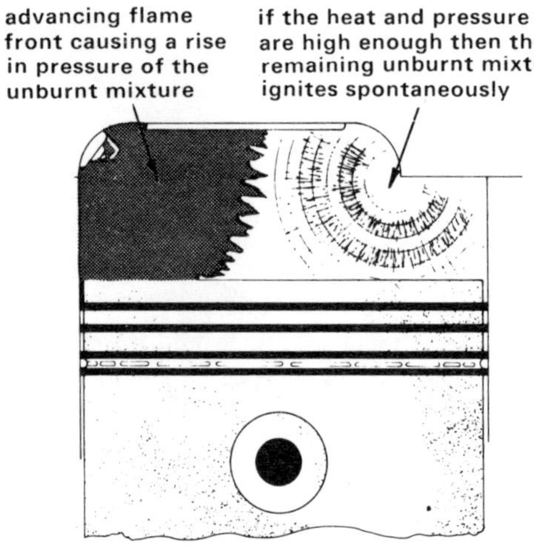

Fig 46 Detonation effect

The effects of high pressure at the time of ignition are a little more complex. Basically it depends on the fact that a high cylinder pressure is prevailing. If the pressure is high because the fresh charge has picked up a lot of heat from the engine, then the charge temperature as well as the pressure, will be high at the point of ignition. This again, will end in detonation occurring if the situation gets out of control. On the other hand if we have a high pressure because the initial charge density was high, thus implying good cylinder filling and at the same time the charge has not picked up too much heat from the engine, then we are on the right road to success. What we have established so far is that an engine which breathes well, will, all other things being equal, have a faster flame front travel than one that does not, but as I have said before, generally speaking temperature, pressure and density factors alone still do not give us enough for complete burning.

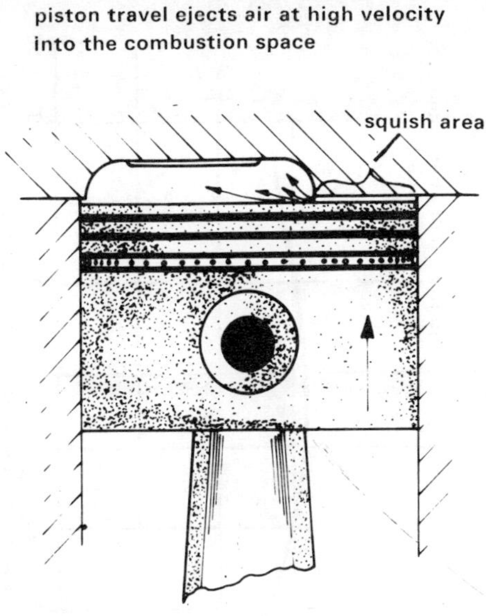

This is where turbulence and swirl come into the picture. If, instead of having the flame front burning through stagnant air, we stir the air into motion so that it goes to meet the flame front, we speed up the rate of burning. The turbulence in the chamber is brought about by the piston coming up the bore and ejecting the air between it and the squish area of the head (Fig. 47). When we relieve the shrouding around the valve, we inevitably cut down on the squish area which may start to reduce the turbulence to the point where burning is too slow. On the other hand, by relieving the shrouding we have enabled the engine to draw in a denser charge of air which, in itself will speed up the rate of burning. In practice we find that the increase in burning rate due to better breathing is far higher than the reduction in burning rate caused by loss of squish area. Because unshrouding the valves shows a diminish-

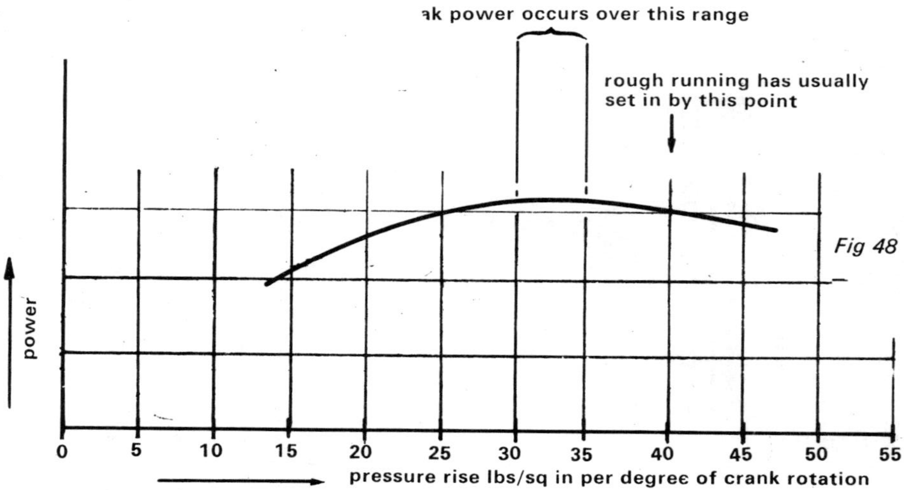

Fig 48 *Optimum pressure rise per degree of crank rotation*

ing rate of return we can, if too much of the chamber is cut away, find ourselves with a burning rate which is not as fast as could be desired.

On the other end of the scale, we can arrive at a situation where, having extracted all the breathing potential possible, we still have too much squish area and our burning rate is too fast. This situation tends to occur mainly in short stroke motors. If the burning rate is too fast, then we could be in a similar position, as regards power, as if the burning rate were too slow.

It has been shown by research into the subject that the best power output from an engine is obtained when the rate of pressure rise is about 30lb per square inch per degree of crank rotation (Fig. 48). If the pressure rise significantly exceeds 35lb per square inch per degree the engine runs decidedly rough. This is because the combustion shock tends to set up vibrations within the engine's components. The figure before rough running occurs will vary a little from engine to engine, depending on how stiff and well damped, are the components of that particular engine.

By designing a chamber which has a slow rate of pressure rise at first and then building up to a high rate of pressure rise, we find that the figure of 35lb per square inch per degree of crank rotation can be considerably exceeded before rough running occurs, but such chambers are not always conducive to high power output. The solution to this problem when it does occur, which is usually rare, is to reduce the effectiveness of the squish area.

Apart from squish-induced turbulence, we can impart a general rotating motion to the air in the chamber by having a curve in the port just before the valve. Because of the shrouded section of the valve in conventional pushrod engines, we tend to get a certain amount of swirl whether we like it or not, so it is not usually necessary to try and build in any more.

When opening out the chamber around the valves, there is one more problem we have to consider and that is the length of flame front travel. The longer the flame front travel is, the more likely we are to get detonation occurring, simply because of the greater length of time the unburnt mixture is exposed to the heat of the burning mixture. A long flame front path also implies that we have a chamber with a large surface area and if this is so, more heat will be conducted away during expansion and lost, thus reducing power. These two points are amongst the primary reasons why a number of small cylinders will develop more power than fewer but larger cylinders of equal volume, again a possible compromise in developing. If we carve too much metal out of the chambers in an effort to gain better breathing, we may find that the flame front path has been increased to the point of detriment, so much so that the CR may have to be much lower than otherwise possible to prevent detonation occurring. Ideally we want the flame front travel to be as short as possible and the flame front should reach all the extremities of the chamber at the same time. Because the flame front travels faster in hot gases, it is usual to have the plug slightly biased towards the inlet valve as the burning time across the exhaust valve is faster than across the inlet valve.

When a chamber profile is altered, the surface area of the chamber inevitably increases. During combustion there is a layer of air and petrol adjacent to the chamber surface which does not burn because too much heat has been conducted away from it by the surfaces in question. This problem gets worse as the surface area to volume ratio of the chamber gets larger. The ideal chamber in this respect is a true sphere, as this has the least surface area to volume ratio. The basic process involved is as follows. When a fresh charge is induced into the cylinder, a certain amount of fuel clings to the chamber's surface. During combustion there is insufficient time for the fuel to heat up and vaporise. After the combustion process is completed, there is a longer period during the remaining phase of the power stroke and the following exhaust stroke when some of the unburnt petrol has time to pick up enough heat to vaporise and is therefore ejected on the exhaust stroke. During the next induction stroke, the chamber surface is re-wetted by petrol taken from the incoming charge. This process repeats itself every cycle and in consequence we find that a certain amount of petrol is wasted. Air is cheap but unfortunately petrol is not, so in the interest of economy we should keep an eye on the surface area to volume ratio of a chamber. Fortunately one of the consequences of raising the CR by skimming the head is that a reduction in surface area is made. This does not mean that any significant change in the surface area to volume ratio is brought about because the chamber volume is also reduced, but the fact that we have a smaller chamber for a given size of engine means that there is less area upon which to deposit fuel. The better the surface area to volume ratio, the less the hydrocarbon emission will be. Low hydrocarbon emission, however, is not necessarily compatible with power output, but a low surface area to volume for the chamber, providing it does not have adverse effects in other quarters, is a consideration for economy. The reason for this is that when maximum power is sought, a rich mixture is used, so the excess fuel ultimately ends up as hydrocarbon emission irrespective of the qualities of the combustion chamber. If the chamber is good, it just means that there would be less than that produced by a bad chamber. Fuel consumption and hydrocarbon

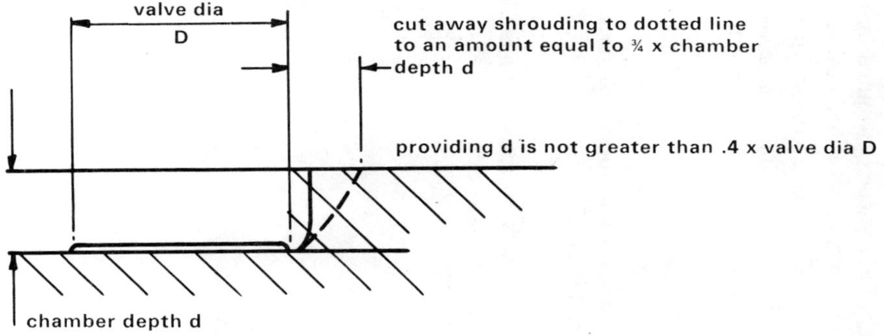

Fig 49A Generalised chamber shape

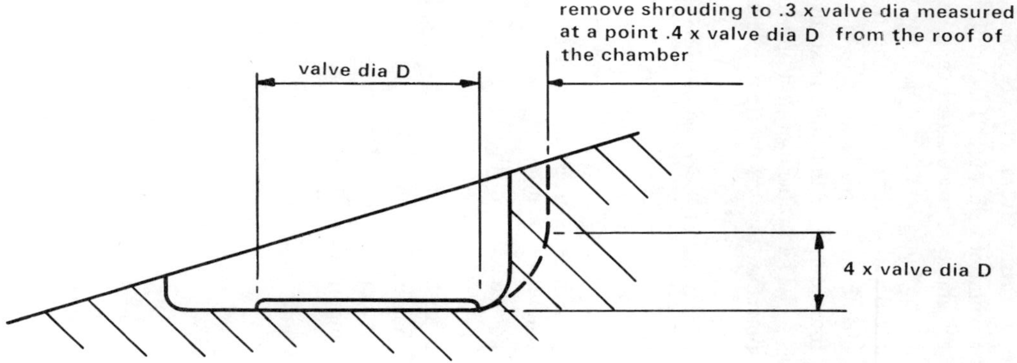

Fig 49B

emission are linked; the more emission an engine has, the more fuel it requires for a given power output, a point to remember when modding heads for road use.

Aside from any considerations of burning characteristics, the amount of reshaping of a chamber will depend on a number of factors; the shape of the original chamber, position of the valves with respect to the bore, whether or not the valves are inclined from the vertical, the depth of the chamber and the port shape leading to and from the chamber. All this leads up to one thing, designing a chamber for maximum output is not quite as straightforward as it might at first seem.

The problem of grinding the chamber to a shape best suited for power is, to a great extent, alleviated by the inclusion of modification drawings within this book, but in case the reader is bent on modifying a head not included, some provision is made for estimating what needs to be done. Remembering all the factors which must be taken into account and disregarding some of the lesser ones, we find that fairly good results can be obtained when the chamber is relieved such that it is three quarters of the depth of the chamber away from the valve seat (see Fig. 49A). This will apply, providing the chamber depth does not exceed .4 of the valve dia. If the chamber significantly exceeds .4 of the valve dia., then we should relieve the chamber .3 of the valve dia. from the seat at a depth of .4 of the valve dia. from the roof of the chamber (see Fig. 49B). If the CR can only be raised a small amount, say, up to about 1 ratio, we should take out a shade less than the amount given above.

chapter nine

Compression Ratio

To many tuners raising the compression ratio is the be all and end all of head tuning. By now I should have made it clear even to the most ardent believers of this theory that compared with breathing potential, the CR is of secondary importance. What I intend to do now is to discuss in detail just how important or unimportant the CR is in its secondary role.

Firstly, when we raise the CR we get more power. The gain in power depends mainly on the amount it has been raised and the figure it was raised from. Under certain conditions we can predict quite accurately just how much more power will be gained by raising the CR a certain amount by applying the following formula:

$$E = 1 - \frac{1}{R^{k-1}}$$

where E = the Otto cycle efficiency

R = the compression ratio

k = the coefficient of adiabetic expansion for air which is 1.4 for true adiabatic expansion.

From this formula, the graph in Fig. 50 has been made up and is far more convenient for the reader to use than the formula. Before applying this graph as a practical means to indicate how much more power we can get by an increase in CR, we must realise its limitations. For instance the figure of 1.4 for true adiabatic compression or expansion of air is not likely to be realised in practice, since there is almost certain to be some heat transference during compression. However, other factors tend to come into play when the CR is increased, which tend to offset any errors there may be in the selection of the value of k.

The big stumbling block associated with the use of the equation or the resultant graph is deciding just how much the CR has been increased. This last statement may leave the reader quite perplexed, especially if he knows full well how to work out the new CR if it has been raised. The truth of the matter is that there are two compression ratios applicable to any one engine; that is the theoretical CR (TCR) and the effective CR (ECR). The TCR is the ratio that is worked out by measuring all the volumes that go to make up the total chamber volume and then applying the formula

$$CR = \frac{V + C}{C}$$

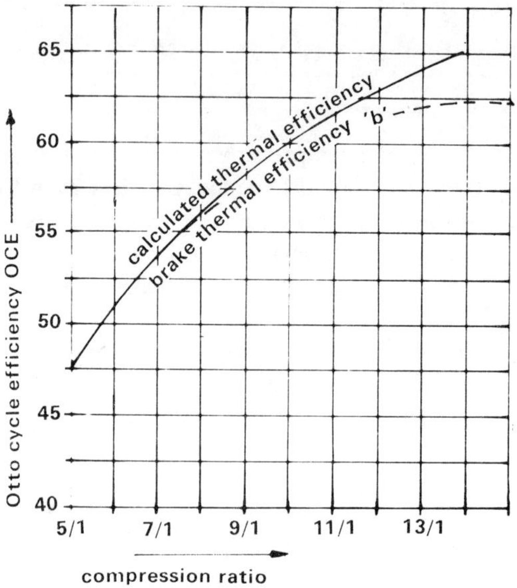

'b' shows the results obtainable from a 100 per cent efficient engine in which the TCR equals the ECR

NB at ECR's above 14/1 the power output drops off, this is due to unfavourable chemical dissociation reaction absorbing the heat. This heat is not lost but it does not become available until too late in the cycle to do useful work.

To get an idea of the amount of power increase use the following formula in conjunction with the graph:

$$\frac{\text{OCE at new CR} \times \text{original bhp}}{\text{OCE at original CR}}$$

Fig 50 Effect of increase in compression ratio on Otto cycle efficiency

Let us apply this formula to a hypothetical example as in Fig. 51A. Here we have a cylinder of 100cc swept volume and a combustion chamber of 10cc volume. By using the above equation the TCR works

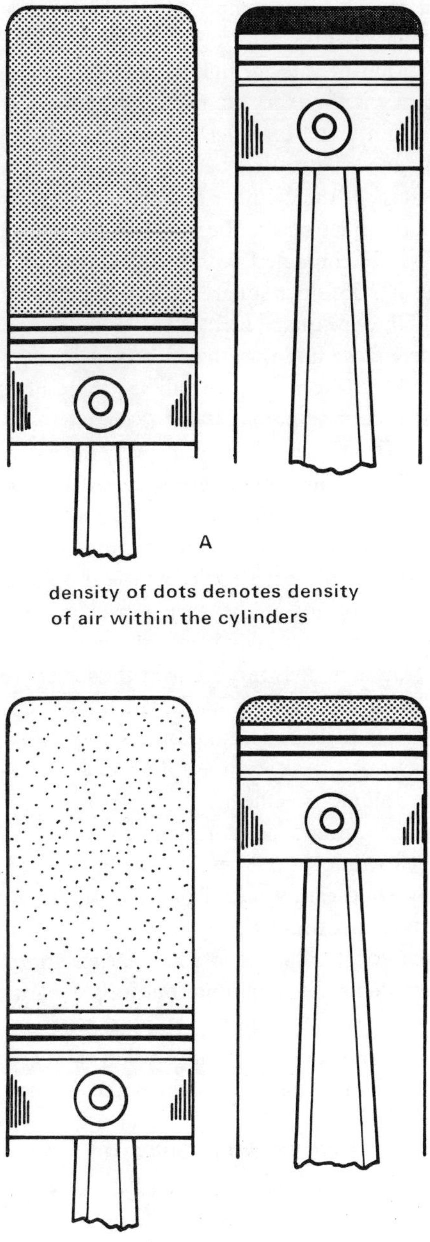

Fig 51

Theoretical and actual compression ratios

out to 11/1 that is, we have compressed 110cc of gas at 14.7lb per square inch (normal air pressure) into 1/11 of its original space.

The implication of this formula is that the engine has a 100% volumetric efficiency ie for every cc displaced by the piston, a cc of air is drawn in. In a simplified form let us consider what happens to the CR when the engine is throttled back so that the cylinder does not draw in a full charge. If the cylinder is only half filled, then (Fig. 51B) at the bottom of the inlet stroke there will only be 7.35lb per square inch of air (14.7 × ½) contained within it. Instead of saying that we have 100cc of air at 7.35lb per square inch, we could say that we have 50cc of air at 14.7lb per square inch ie the cylinder is half full. If we now work out our CR we find that the effective CR is 6/1 although our theoretical CR is 11/1. From this we can see that our effective CR is dependent on the breathing potential of the engine.

In practice an engine will rarely, if ever, have a volumetric efficiency of 100%. The natural restrictions caused by air filters, carbs, ports, valves and exhaust systems all limit the volumetric efficiency. As engine revs rise, so volumetric efficiency drops off. When we reach the point at which the volumetric efficiency is dropping off as fast as the engine revs rise, we are at the peak power point of the engine. Because all modern engines employ camshafts which have overlap, the peak volumetric efficiency does not occur at the lowest revs the engine will run at. The longer the cam timing is, the further up the rev range the peak volumetric efficiency is, and this is the point at which the engine will put out its peak torque. Because our effective CR is dependent on the volumetric efficiency, the greatest per cent gains in power will be at and around peak torque rpm. For instance, raising the CR by, say, two ratios, could well give a gain of 5bhp at about 3,000rpm but only 1bhp gain at 5,500rpm. At this stage it should now be clear why a good breathing ability is important to an engine, for without it our theoretical CR is, in practice, never approached.

When we are going for maximum output we want the CR to be as high as possible within the limits set by detonation. The factors which affect our maximum useable compression ratio are as follows:

1. The better the breathing capabilities the engine has, the lower the maximum CR.

2. The hotter the engine temperature, the lower the maximum CR.

3. The shorter the flame front travel, the higher the maximum CR.

4. The greater the anti-detonation properties of the chamber, the higher the maximum CR.

5. High octane fuels and rich mixtures increase the maximum CR.

6. Camshafts which employ longer duration and overlap can be run with a higher maximum CR than ones having short timing and overlap.

Taking these points step by step, we have seen how the breathing potential affects our maximum CR. As our volumetric efficiency gets nearer to 100%, so the ECR gets nearer to the TCR. As an example, if we have an engine which at 100% volumetric efficiency can accept a 12/1 CR maximum, this very same engine would, at 50% volumetric efficiency accept a maximum CR of roughly 24/1.

The second point in the list should, by what has been said earlier in this chapter, be self evident. As the temperature of the mixture rises, so it is bound to be more prone to spontaneous self-ignition. Under this heading we can include detonation caused by hot spots in the chamber, such as sharp edges, glowing carbon deposits and plugs which are overheating because they are operating above their intended heat range.

Again point three has been covered earlier in this chapter, but to refresh your memory, the shorter the flame front travel is, the less time the unburnt mixture has to absorb the heat from the advancing flame front.

Point four is one thing that we have not yet covered except by the implications of point three. In the section dealing with port design, we discussed at some length the advantages of curved chamber walls in and around the vicinity of the valves. The use of the correct curve on the chamber walls results in the least chamber surface area compatible to good breathing. Also stagnant burnt gas from the previous cycle with its resultant heating and dilution of the incoming charge are more likely to be minimised.

Point five—high octane fuels are by their very nature meant to postpone the onset of detonation. This is a commonly known fact. Not so obvious is the fact that a rich mixture is less detonation prone than a chemically correct one for complete burning, or a weak one. The underlying reason for this is that the excess fuel cools the incoming charge and the combustion chamber surfaces. For road use, excessively rich mixtures should be avoided since they do have a noticeably adverse effect on economy.

The final point—camshaft timing. How can this affect our maximum CR? Basically it all boils down to volumetric efficiency and its relation to the ECR. If we have two camshafts of differing timing, then assuming the timing is correct to make full use of the burning mixture on the power stroke, the cam which gives the most power is the one which will allow the engine to pass the greatest volume of air and petrol per minute, and not, as one would at first think, the one which gives the highest volumetric efficiency. This might, at first, seem a rather ambiguous statement, but not so if we consider that horse power is the rate at which an engine is capable of doing work. From this it follows that our hp must be proportional to the rate at which the engine consumes air and petrol, not the efficiency with which it can draw in air in one cycle. An engine which has a volumetric efficiency of about 60 per cent at 10,000rpm is consuming air and fuel at a greater rate than an engine which is 70 per cent efficient at 7,000rpm. The ratio at which the first is better than the second is

$$\frac{0.6 \times 10{,}000 \times 100}{0.7 \times 7{,}000} - 100 \text{ per cent} = 22 \text{ per cent}$$

Because the volumetric efficiency of the high lift long duration cam can be less than the one having shorter timing, we can use a higher theoretical CR so that our effective CR is the same as the second case.

If the engine in question is employing shock wave and inertia tuning of the inlet and exhaust, the volumetric efficiency at one particular point may be increased and as a result, the maximum CR will have to come down slightly, usually in the order of half a ratio. As a general safe rule, the maximum CR with a full race cam is in the order of 1½–2 ratios higher than the maximum CR with a road cam.

chapter ten

Optimum Valve Sizes

When modifying an engine, it is essential that it is 'purpose tuned', that is, due regard is given to the purpose for which the engine is to be used. Coupled to this, the engine should have the characteristics most desired by the driver. For instance, it is not really a good idea to have an engine which produces its peak torque at 5,000rpm and peak power at 7,000rpm fitted in a car that is to be used for towing a caravan. Such a combination would be extremely tiresome to drive as well as having a nasty tendency to wear out clutches quickly.

The only answer is to produce an engine which is best suited for the job in hand and nine times out of ten we can look at this problem solely from the point of view of the rpm at which peak torque and peak power occur. If the engine is expected to perform heavy work such as towing, pull high gears at low engine revs and be flexible, then the peak torque must occur fairly low down in the rev range. If a sporting performance is required for fast road driving, then the point at which peak torque and peak power occurs can without significant adverse effects, be moved up the rev range.

When all out power is required, such as would be the case for a racing engine, much has to be sacrificed. Low down and mid range torque, smooth running and flexibility are inevitably lost with very high specific output engines. Basically it all boils down to one thing, that we cannot get the best output from an engine at both ends of the rev range, and so we must be sure that the power occurs at the most used point in the rev range.

One factor that contributes greatly to controlling the point at which peak power occurs is the size of valves employed. The bigger the valves, the further up the rev range peak power occurs. Generally speaking, the peak power occurs on a racing engine, as opposed to a modified production engine, when the gas speed through the valve orifice reaches a theoretical average speed of 250–280ft/sec. The formula for working out the theoretical gas speed is as follows:

$\frac{Cv}{Va} \times \frac{rpm}{360}$ where Cv is the capacity in cubic inches of one cylinder and Va is the area of the inlet valve in square inches. Before we can use this formula, a good understanding as to why peak power should occur at a specified theoretical gas speed is required, otherwise we shall be no better off.

Let us assume that we have an engine whose volumetric efficiency (VE) is 100% (or any other per cent so long as it is unchanging) no matter what rpm it is turning at. Such an engine would give more power the higher it was revved. Its peak power would occur at infinitely high revs, the power output itself would be infinite and the gas speed at which peak power occurred would be infinite. This state of affairs cannot, unfortunately, be realised in practice. Instead, what happens is that the volumetric efficiency drops off as revs rise. At peak power revs, we find that our VE is dropping as fast as our revs increase and that is why we get no more power when revs rise beyond this point. The main governing factor controlling our VE is the pressure drop within the inlet passage, or to look at it another way, VE is the ratio between normal atmospheric pressure and the pressure that exists within the cylinder when it has filled as much as it is going to at the particular revs concerned. This means that if a pressure of 7.35lb per square inch exists in the cylinder when the inlet valve closes, then the VE is

$$\frac{7.35}{14.7} \times 100 = 50\%$$

The slower the rate of pressure drop between the cylinder and outside atmosphere in relation to the rate at which revs rise, the higher theoretical gas speed our peak power occurs at. Gas flowing tends to increase overall volumetric efficiency while high lift long duration camshafts tend to delay the point at which volumetric efficiency drops. Such camshafts in general do not increase the volumetric efficiency, not unless they replace a cam of none too good a design. Such a state of affairs means that in practice we must, when trying to decide on what valve size to use, take into account the camshaft timing. With pushrod engines the situation is further complicated by the fact that we have, to a varying degree, valve shrouding. This has the effect of lowering the gas speed at which peak power occurs, and siamese ports lower the figure even further. With so many variables to contend with, we find that it is possible to predict only the approximate point at which peak power occurs. Taking into account the shrouding caused in most pushrod engines, we find, in general, that peak power occurs at approximately 230–240ft/sec for an engine in race tune. For a pushrod siamese port racing engine, the figure is normally 200–220ft/sec. On top of allowances made for head design, we must take into account the effects of the cam timing. Lengthening the timing will raise the figure and shortening the timing will lower it. In order to account for the effects of cam timing, I have found the following figures very useful for estimating the gas speed at which peak power will occur.

Cam	Siamese Ports	One port/valve	Inclined valves
Full race	200–220	230–240	240–280
Half race	170–180	180–190	195–210
Road	150–170	.160–180	175–195

Once the revs at which peak power is required has been decided, and remember it must bear some relation to type of cam, a road cam won't let an engine breathe at 10,000 revs no matter how big the valves, we can work out the valve dia. from the following formula:

$$d = \sqrt{\frac{CV \times rpm}{\pi \times GS \times 90}}$$

where d = valve dia. in inches
CV = volume of one cylinder in cubic inches
π = 3.142
GS = gas speed

Using the above figures, the inlet valve size can usually be estimated within 10% which at least gives a starting point. One thing that we must watch with pushrod engines is over valving. It is always better to err on the small side unless you have plenty of spare heads to experiment with.

Once the required inlet valve size has been established, the selection of an exhaust valve size becomes simple. The general practice is to make the exhaust valve between .80 and .90 of the dia. of the inlet valve. The reason why the exhaust valve does not need to be as large as the inlet is because the gases are forcibly expelled by the piston whereas the inlet has to draw in air, with the minimum drop in pressure, from a pressure source which is not very high to start with. Unless the engine is supercharged, the largest pressure pushing air into the cylinder is 14.7lb per square inch. On the exhaust side there is generally 70–80lb per square inch in the cylinder at the time the exhaust valve opens. This, plus the fact that the piston displaces the exhaust, means that spent gases are quite positively expelled. By increasing the exhaust valve size, we can reduce pumping losses but the exhaust valve size must never be increased at the expense of the inlet valve size.

When selecting valve sizes for a racing engine, we do tend to find, on many occasions, that all the available room for valves is required, ie the diameter of the inlet and exhaust together are nearly equal to the bore size. Under these conditions the ratio of inlet size to exhaust valve size becomes more critical, especially if we are to extract the maximum power. The most important point affecting this ratio under these conditions is the efficiency of the inlet port and valve. The more inefficient the inlet port and valve are, the greater the inlet valve size will be relative to the exhaust valve. To illustrate the point, let us compare the inlet to exhaust ratio, each one with a more efficient inlet port and valve than the one preceding.

Push rod siamesed port	inlet to exhaust ratio 1–.8
Push rod one port/valve	inlet to exhaust ratio 1–.85
Twin cam hemispherical head	inlet to exhaust ratio 1–.9

When supercharging an engine, we considerably increase the ease with which the cylinder is filled. This can be likened to an increase in induction efficiency as it results in a greater amount of air finding its way into the cylinders. Under such conditions we have to reallocate the amount of room given to each valve. With average boost pressures, 7–10lb per square inch, the inlet to exhaust ratio should go up about 0.05in on the previous figures, that is 1/.8 becomes 1/.85 and so on. When very high boost pressures are used on an engine with inherently good breathing characteristics, we can end up with the inlet and exhaust valves being virtually the same size.

When ample room does exist to accommodate the valves likely to be required, then selection of the exhaust valve size is not so critical but we still must avoid over valving. If the exhaust valve is made too large it will, more often than not, cause the low and mid range torque to suffer slightly, especially if the exhaust port size matches the valve size. The best plan is to stick fairly closely to the inlet size to exhaust size ratios given earlier.

chapter eleven

Valve Springs

The sole function of the valve spring is to control the motion of the valve and its associated valve train such that it follows the motion prescribed by the cam profile to return the valve to its closed position. As such, it must perform this function up to the maximum useful rpm of the engine.

The higher the engine revs are, the stronger the valve springs need to be. Unfortunately as valve spring pressures increase, it is accompanied by an increase in the friction level of the valve train as a whole. This being the case, we must avoid using valve springs that are any stronger than necessary, otherwise it will be paid for in terms of lost power. On an average 1–1½ litre engine, it is not out of the question to lose 2bhp by having valve springs which allow the engine to rev 1,000rpm more than is necessary. In general, the valve spring strength should be such as to allow revs between 800–1,000 revs higher than the rpm at which peak power occurs. In many instances, when a modified head is the only change on an engine, the peak power will occur at more or less the same revs as the standard engine. Under such conditions it is not a good idea to fit stronger valve springs because the extra spring pressure will serve no useful purpose.

When peak power starts to move up the rev range, then some thought should be given to extra valve spring pressure. If, for the moment, we ignore any possible reductions that can be made in the reciprocating weight, the biggest factor affecting the valve spring strength is the camshaft. This is so for two reasons (a) the camshaft plays an important part in dictating how far up the rev range peak power will occur and (b) the acceleration and deceleration which the valve spring is expected to control is imparted by the camshaft in the first place. From this we can see that the camshaft plays a major role in deciding the valve spring strength required. This being the case, we should, strictly speaking, select our valve springs to match the camshaft.

The ideal springs are those which are the weakest that will do the job required. We can relieve the spring of some of the work required of it by lightening the valve, the spring retainer, rocker, pushrod and cam follower. If this is done, then we should, in theory, if the valve springs were of sufficient strength to allow the desired rpm in the first instance, reduce the spring pressure so that valve bounce rpm stays the same. In practice we find that the reduction of the overall reciprocating mass that can be achieved is small, and a figure of 10% for the overall reduction in weight would represent the results of a fairly intensive lightening programme.

With the valve gear so lightened, we can either reduce the valve spring pressure by a corresponding amount or accept the increased rpm before valve bounce occurs. The 'extra' rpm is sometimes quite useful especially on a track car. The situation often arises when travelling between two corners which are close together, when it is impractical to change up because no sooner had the upward change been made, a downward change would be required for the approaching corner.

When fitting valve springs there are two things which we should watch: (a) that the valve springs are all the same fitted height and (b) at the fitted height the pressures exerted are the same. Checking the fitted height can be done quite simply as shown in Fig. 52, that is

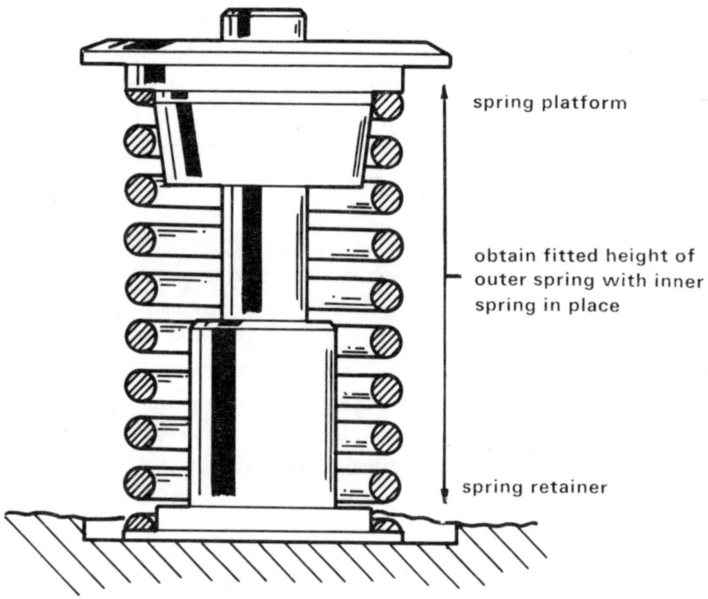

Fig 52 Checking fitted height of valve springs

by fitting the inner spring and measuring from the spring platform to the relevant face of the spring retainer.

Checking the pressure exerted when fitted is a little more cumbersome and one of the simplest ways to do it is as shown in Fig. 53. When using this method the force required to just open a valve will be equal to $\frac{R_1}{R_2} \times$ Wlb. When fitting valve springs in this manner, we should endeavour to get the installed heights of each spring within 0.010in and the pressures extended within 2lb. Small errors in pressure

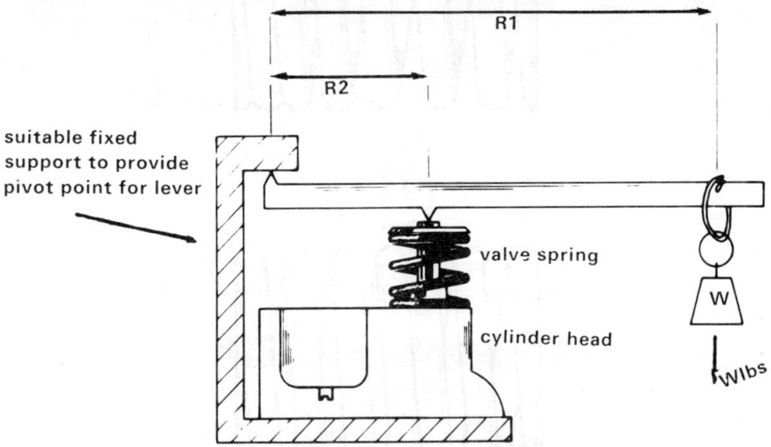

Fig 53 Testing valve spring pressure

and height can be rectified by shimming the spring platform. Errors other than small ones will require a change of spring. For road-going engines, setting the spring pressure to very fine limits is not worthwhile. On racing engines where competition is close and every bit of power helps, it becomes more essential.

The advances in metallurgy and design have, over the years, brought springs to the point where they are almost 100% reliable. Valve spring breakage is indeed quite a rare occurrence. Generally speaking there are three points which can give rise to spring failure, these being a flaw in the metal, incorrect tempering and spring surge. The first two points, spring manufacturers go to great lengths to control, so failures from these quarters are minimised. Spring surge, however, is something which occurs on the engine and is not necessarily inherent in the spring.

Spring surge is caused by the spring vibrating at its own natural resonant frequency which it will do if it is excited by a periodic force

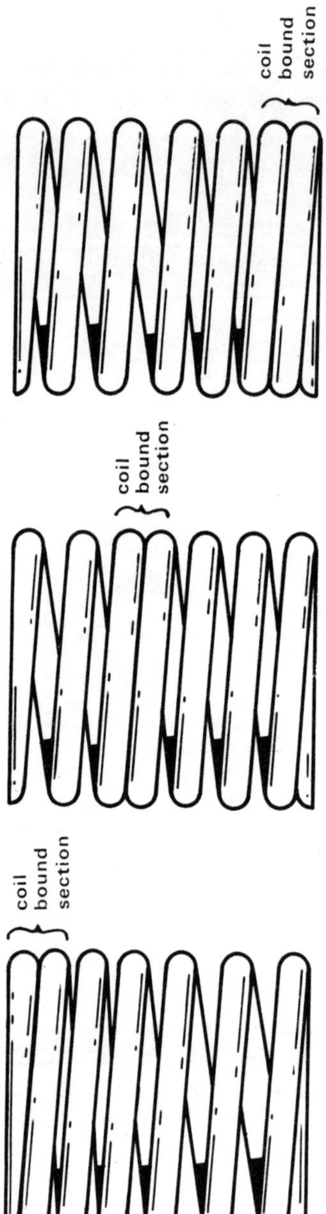

Fig 54 Spring surge causing coil binding

whose frequency is at or around the natural frequency of the spring or some multiple of it (Fig. 54). On normal road engines we find that the natural frequency of the spring is appreciably above the frequency of any primary periodic forces or motions such as imparted by the camshaft and multiples of the periodic force tend to be of diminishing amplitude, so these are not troublesome to any extent. As spring pressures increase and engine revs rise, so the possibility of spring surge becomes more real. To prevent the spring surge reaching excessive proportions, we must introduce friction damping in one form or another. This is usually achieved by using a double or triple spring set-up instead of a single spring. If the springs are a close fit one within another and wound in opposite directions and the fitted lengths differ, they will, during operation, rub together, thus providing a damping effect on any unwanted vibrations that may be imminent. Bearing this in mind, it is better to use double or even triple springs to achieve the required spring pressure rather than a single spring of equal strength.

chapter twelve

Spark Plugs

Once an engine has been modified, it may be found that the original plugs are no longer suitable, but we can, by using the wrong plug, suffer poor performance or even severe engine damage. The use of the correct type of plug is therefore of prime importance. The ability to 'read' a plug or take a 'plug count' as it is sometimes called, is an invaluable asset to the engine tuner as it enables him to get a fair idea of the conditions prevailing in the combustion chamber.

The photographs in Fig. 55 show the most common maladies found by plug inspection. Upon these we must superimpose the conditions brought about by the fuel/air ratio and ignition timing. Assuming at this stage the right heat range of plug is being used, a weak mixture will cause the plug to take on a light brown or tan colour. Such a state of affairs can lead to damage to the engine if allowed to persist, especially if the engine is driven hard. The most likely things to happen if a weak mixture is used is that the exhaust valves burn out, but quite apart from this, a weak mixture can cause a considerable reduction in power output. A black sooted plug indicates that the reverse situation is prevailing, that is, the mixture is far too rich. Although infinitely preferable to a weak mixture an excessively rich mixture will cause loss of power, high fuel consumption and an increased rate of bore and guide wear. If the plug appears black, but not sooty, with a trace of brown, especially around the insulator nose of the centre electrode, the mixture is probably about right for optimum power.

The state of ignition advance has an appreciable effect on plug temperature. Advancing the ignition from the optimum setting will cause the plug temperature to rise, and in doing so bring about the conditions for pre-ignition. When the ignition is over advanced, the plug temperature can, in extreme cases, rise from 600°C to as much as 1,100°C. The diagram in Fig. 56 indicates the ideal operating range for plugs, and when choosing a plug for a particular application, we must try to pick a type which satisfies these thermal requirements.

Since it is imperative that the spark plugs operate in the correct temperature region, we had better look at the controlling factors

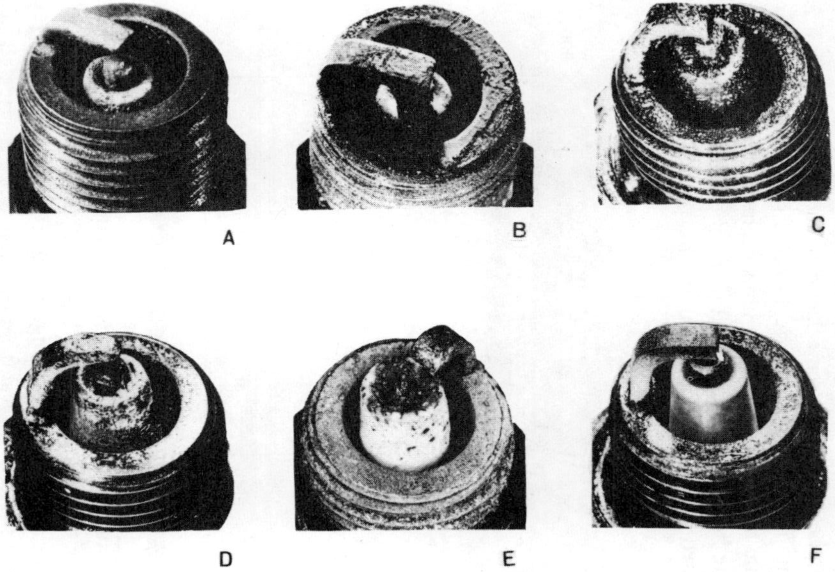

Fig 55 Plug conditions

A *Worn out—replace*

B *Carbon fouling—hotter grade needed*

C *Oil fouling—worn inlet valve or rings —new engine!*

D *Electrodes burnt away—over advanced ignition or weak mixture*

E *Pre-ignition—decoke overdue*

F *Perfect condition*

involved. Firstly an internal combustion engine is a device for transforming heat energy into mechanical energy and as such it is not very efficient. This means that a great deal of heat energy which we hope to turn into mechanical energy is not so converted and must be dissipated as waste heat. Assuming a fixed thermal efficiency, the amount of heat dissipated will be proportional to the power output of the engine.

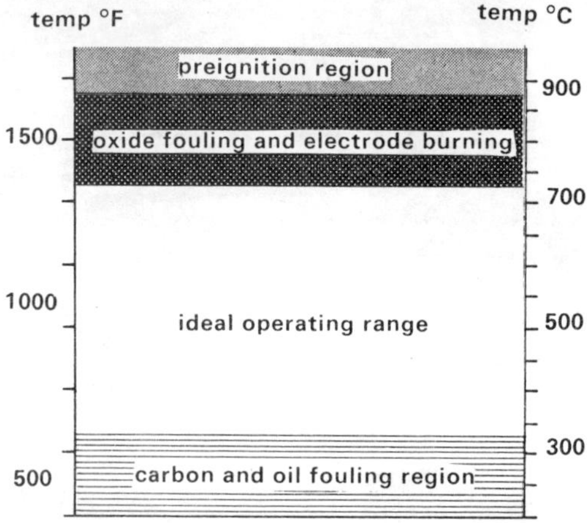

Fig 56 Plug temperature range and effects

Also assuming all other things to be equal, a small engine will run at a higher temperature than a big one if the power output is the same. This means on a small, highly tuned engine, we must pay closer attention to heat removal from the vital areas. Bearing all this in mind, what would happen if we tried to use the same type of plug in (a) a highly tuned small engine, and (b) a large slow running engine?

If the plug was right for the large engine where prevailing temperatures are low and there is plenty of surface area around the plug to dissipate the heat, the plug would suffer severe over-heating when fitted in the small engine. On the other hand, if the thermal conductivity of the plug was high so as to satisfy the requirements of the small engine, it would never reach a satisfactory operating temperature in the large engine. Such a state of affairs would lead to fouling of the plugs because the heat is insufficient to burn off deposits which cause the plug to electrically short out. From this we can see that plug selection is directly affected by temperature requirements, and temperature, within reasonable limits, depends upon the state of tune of the engine.

What governs the heat range or thermal conductivity of a plug?

The simple answer is the length of its heat path, (Fig. 57). If the centre insulator physically contacts the outer shell of the plug a short way from the electrode tip, it is said to have a short heat path, and as such will conduct heat away much quicker than one having a long heat path. The net result then is that as the state of tune of an engine goes up, so the temperatures of combustion rise, thus requiring a colder, shorter heat path, running plug to remain within the ideal temperature range for proper functioning.

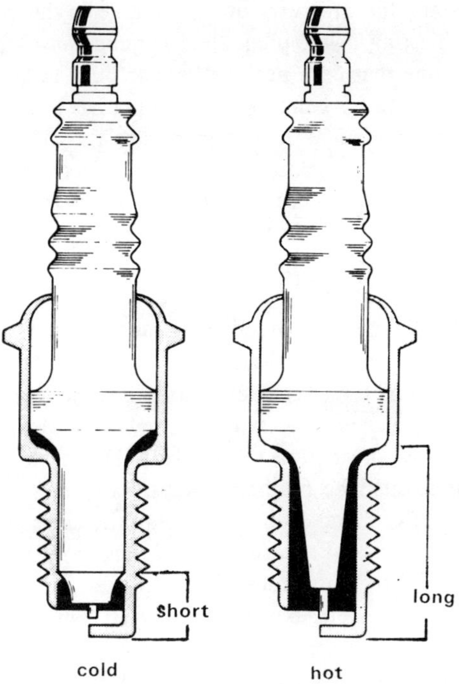

Fig 57 Different lengths of heat paths

At this stage it should be clear what happens if a plug is overheated because of inadequate thermal conductivity to stay within its working temperature range, ie too hot a plug for prevailing conditions. It would, however, be useful to further investigate the other extreme, that is a plug which is too cold for the job in hand.

If the plug is too cold for a particular engine, it will foul up mainly with oil and carbon deposits. If the correct temperature had been reached, these deposits would have burned off. If these deposits do not burn off, they can collect and create an alternative route for the electrical energy that was supposed to create a spark across the plug gap, hence a misfire occurs. The misfire will lead to further deposits on the plug as well as reducing its temperature. This makes the likelihood of another misfire even greater and the whole thing is a vicious circle.

To select a heat range of plug for a tuned engine, one should always start off with a plug that is known to be too cold. Too cold a plug will obviously foul up. From this point you progress up the heat range until you arrive at the first plug which does not foul up and that is the one to settle for. It is very unwise to reverse this process of plug selection, ie selecting a hot plug, then coming down the heat range until you find one that does not overheat or burn away. Damage due to pre-ignition may arise before the right plug has been found.

Apart from variations in design due to thermal considerations, we find other features introduced notably in electrode design, which enhance, or in some cases dictate, the use of a particular plug for a particular application. The first such feature to spring to mind is the extended nose plug of which a Champion N9Y or N6Y is a typical example. The extended nose plug has a very wide operating range inasmuch as it will function correctly over ranges which would normally be covered by three or four ordinary plugs. Because the extended nose of the plug positions the electrodes deeper into the chamber, the plug picks up heat quicker and easier during low speed operation such as town driving, thus preventing any fouling. During high speed wide throttle requirements, the plug temperature is prevented from rising too high by virtue of being deep enough into the chamber to be cooled by the incoming charge. Because of its wide heat range, the extended nose plug is eminently suitable for tuned engines.

Another type of plug is the auxiliary air gap plug. Apart from the normal gap at the business end, this plug has a secondary air gap between the HT connector and the centre electrode. The purpose of this air gap is to cause a faster voltage rise time at the firing electrodes or principal air gap. The faster rise time at the spark gap means there is less time for the current to drain away because of fouling. This enables a fouled up plug to still give a spark. Such plugs are normally used in slow running stationary engines, ie pumps, electricity generating sets, etc., where the temperature never reaches a satisfactory level for a more conventional plug.

A third variation on the standard type of plug is the sort which has the side electrode and the centre electrode within the confines of the outer shell. This type of plug is commonly used for racing and its design features allow for two things. The use of high compression raised crown pistons may mean that the pistons can strike the electrodes of an ordinary type plug, hence the use of retracted electrodes on the racing type. Secondly, heat conduction from the tip of the electrode is slightly superior so the possibility of pre-ignition is reduced. The heat range of racing plugs is such as to make them ideal for flat-out driving with highly tuned engines, but prone to fouling at tickover and low speed conditions.

If a fuel other than petrol is used, then it may be found necessary to make a plug change to suit the fuel used. If the sole change of an engine is to run it on methanol, with suitable adjustments to jetting, instead of petrol, then a plug one heat range colder will usually do. On the other hand, if the CR is increased when the change to methanol is made, possibly the plugs will need to go colder still. If nitromethane is being used, a very cold running plug will be needed because nitromethane becomes unstable at high temperatures and pressures.

To enable the reader to more readily see the difference between heat ranges of a given type of plug, a Champion specification chart is reproduced in Appendix A.

PART THREE

Head Designs and Applications

So far we have looked at the general techniques to apply when modifying a head, but because we are faced with so many variables, we must in the final analysis, tailor the head to our requirements on a flow machine and an engine dynamometer. If the reader works along the lines put forward in an earlier part of the book, the cylinder heads so produced will be good, but not necessarily quite as good as possible. It is for this reason that the book includes drawings of a number of modified heads which have been dynamometer tested. In doing so we have acquired the final edge so to speak, and gains of the odd few horse power have been realised.

The value of dynamometer testing is not stressed often enough. When the reader has modified a head he will rightfully expect the car car to perform better. If the engine is set up by ear after having a modified head fitted, or a modified anything for that matter, we cannot expect to get the full power potential from the engine. It is possible, when setting up an engine by ear, to get less performance after fitting a perfectly good modified head. This can happen, for instance, if the plug leads are the carbon string type and unknown to the proud owner of the car, they have broken down and their electrical resistance is very high. Under such conditions there may be just enough voltage at the plugs to fire them with a standard CR, breathing ability and engine rpm. When a modified head is fitted, we may have as much as 30% more breathing, a CR 2–3 ratios higher and an rpm potential of 1,000 revs or more higher. With the higher cylinder pressure we find the plugs are no longer able to fire the mixture efficiently; result—less performance than before. If the engine is set up on a dynamometer in conjunction with an electronic engine tuner such as Crypton, Sun or any other type, these problems of minor malfunctions are easily detected and overcome, and the full power potential of the engine realised.

One further point which bears mentioning, tuning should never ever be regarded as a substitute for reconditioning. The best engines to tune are those that are in perfect condition. Modifying engines which are below par can result in a disappointing performance and unnecessary added expense.

The first stage for 848 & 948 BL engines

It is almost inevitable that we start this part of the book with something suitable for Mini engines, but apart from Minis, the head we are about to deal with is suitable for 948cc engines which have the 1.093in dia. inlet valve. The earlier 803cc engines can use this head but it is best to change the existing head for a later 850 Mini or inline 948cc. One word of warning, however, we should not tune 803 engines too much, as the crank is not particularly strong.

Basically the head should be modified as shown in all the drawings which make up Fig. 59. The only thing that we have to watch with these heads is the amount which can safely be skimmed off. If you look on the rocker pedestal mounting face about number one cylinder, ie the radiator end of the head, you will see an oil hole drilling. This drilling goes through the cylinder head almost to the opposite face and it is generally this hole which limits the amount which can be

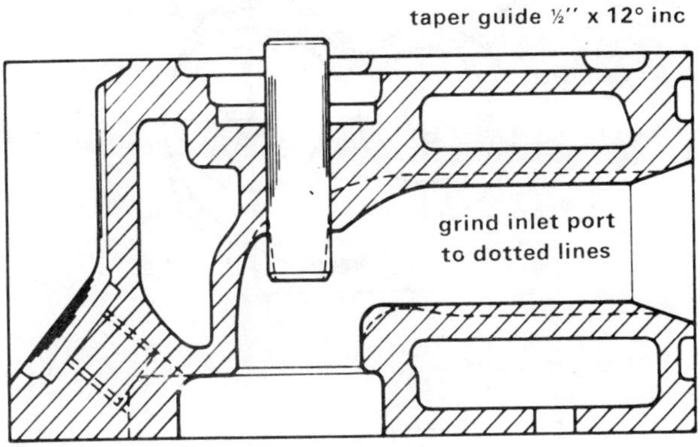

850 Mini port

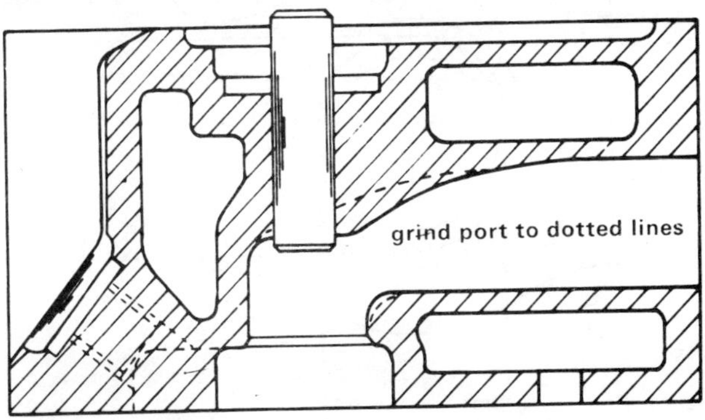

850 Mini exhaust port

*Drawings marked * are reproduced full size in Appendix B.*

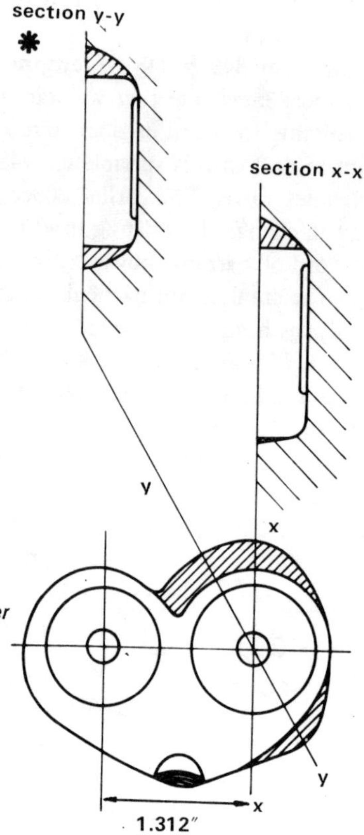

Fig 59 Chamber shape for 850 Mini

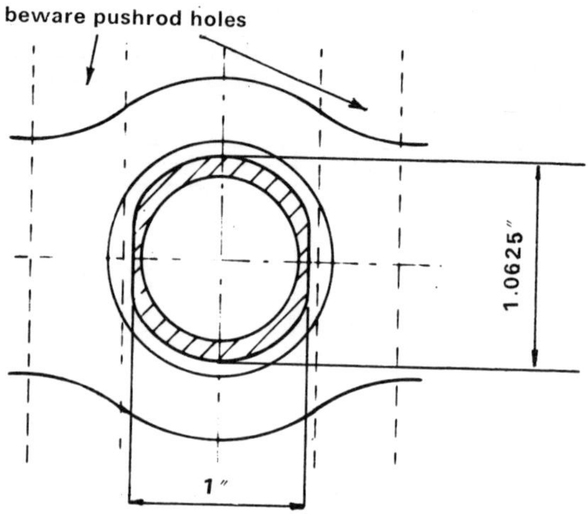

View down inlet port

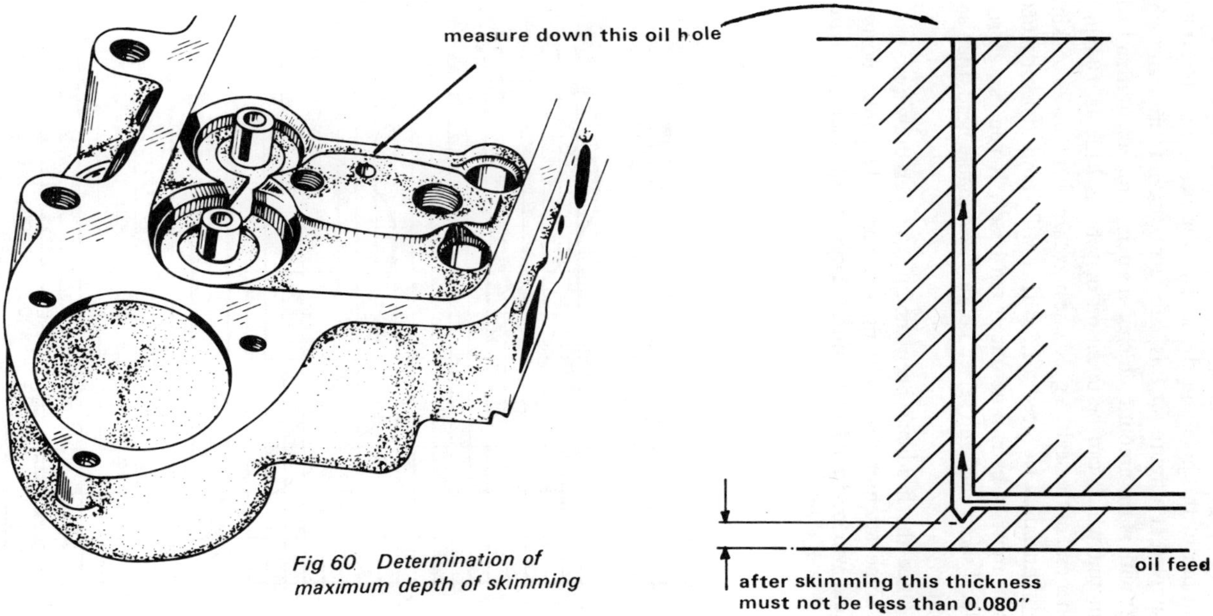

Fig 60 Determination of maximum depth of skimming

skimmed off the head. To determine how much can be skimmed off, measure the depth of the hole to find out how thick the metal is between the bottom of the hole and the face of the head (Fig. 60). Once you have arrived at a figure for this, subtract 0.080in from it and what you are left with is the amount which can safely come off. With a bad head, the amount which can be taken off may be as little as 0.045in and at the other end of the scale, a good one may take over 0.090in of skimming. For our purposes we need a head that will stand at least 0.060in off which in most cases is nine out of ten heads. If you can go round to your local scrap yard and select a highly skimmable head, then so much the better.

Once the head has been modified as per drawings, we should make some attempt to match up the ports. The inlet ports in the manifold will be found to require enlarging by about $\frac{1}{8}$in–$\frac{3}{16}$in at the point it mates with the head.

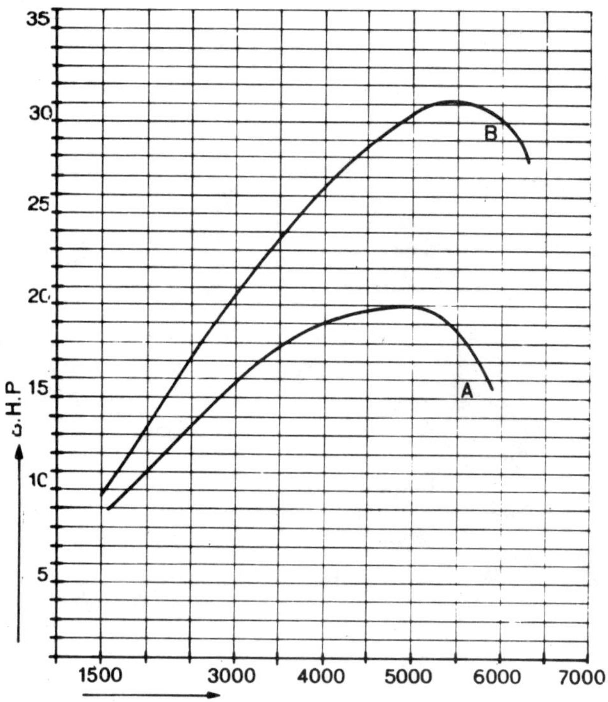

Fig 61 Bhp at driving wheels of 850 Mini

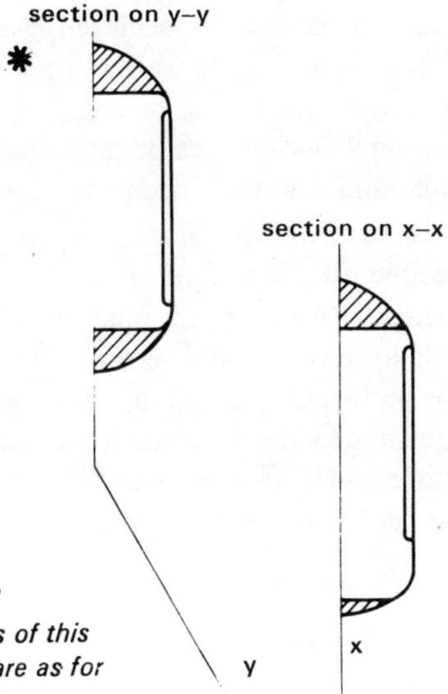

Fig 62
Details of this head are as for Fig 59

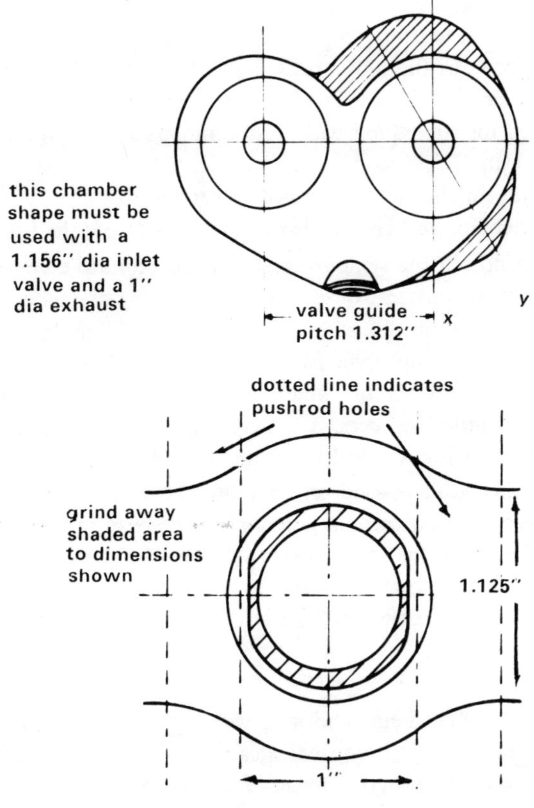

The results that can be achieved with this head are quite spectacular. On an 850cc engine we can, by solely fitting this head and adjusting carburation and timing to suit, put the power output up by some 35%, and on 948cc engines by 30%. If the air filter is removed, a ram pipe fitted and a suitable needle used, the power output can be pushed up even further. The graph, Fig. 61, shows just how much increase is possible using this type of head together with an E3 needle and an exponential curved 3in ram pipe on an 850cc Mini engine. The power output is up by over 40% of that from a standard engine. The power tests for both standard and modified versions were done on the same engine and under the same environmental conditions so a comparison of before and after figures is direct. The use of this head is ideal when the head is virtually the only modification contemplated.

The second step for 848 and 948 engines and the first for 1100's and 997 Coopers

With 850cc and 948cc engines we can, if suitable changes are made to the carburation, usefully employ a slightly larger inlet valve without losing any torque in the lower rev range when compared to the head previously described. To get good results, the single 1¼in SU should be substituted for something with a little more air flow potential such as a single 1½in SU twin 1¼in SU's, a 28/32 Weber or something similar. To do this head you will need a casting that can be skimmed by 0.075in–0.080in. If a single 1½in SU is to be used, then it is quite in order to grind out the standard cast iron manifold to take the larger carb. The inlet valve size to use is 1.156in (1$\frac{5}{32}$in) dia. as opposed to the 1.093in (1$\frac{3}{32}$in) dia. found standard in Minis. This larger size of valve is used in 997, not 998, Mini Coopers and Austin and Morris 1100's. Since the subject of valve sizes has brought us very nicely around to these other two cars, it can conveniently be pointed out that the head shown in Fig. 62 is also suitable for 997 Coopers and 1100's.

When this head is used on an 1100, the single 1¼in carb quite seriously restricts the breathing potential of the head in the higher rev range. To cut down this restriction, the fitment of 1½in SU is a good idea. This head, together with the 1½in SU will give, on a perfect 1100cc engine, an increase of 14bhp at the wheels.

The 997 Cooper being endowed with twin 1¼in SUs as standard, need not have any carburation changes made, except re-needling and re-setting, to obtain very good results from this head.

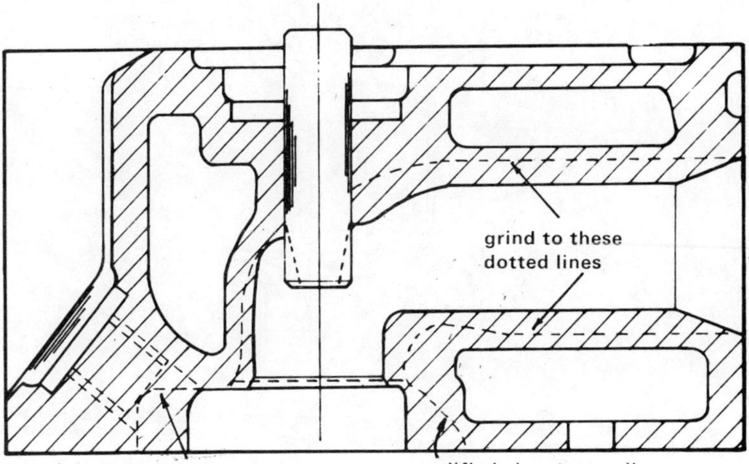

Big inlet valve head for 848, 997 & 1100 engines

The use of this head shown in Fig. 63 is eminently suitable on 850cc engines where a change in cam has been made. It was designed originally to give the best results with the excellent BLMH camshaft C-AEA-731. It does, however, work very well on the 997 Cooper and 1100 engines without a change of cam but the use of the aforementioned cam on either of these engines with this head is really worthwhile. The design of the head is such as to give low end pulling power from 1,000 rpm upwards even on 850cc engines. The graph, Fig. 64, shows the power output of a standard 850cc engine compared with the power output of an 850cc engine modified by using (a) the head if Fig. 63, (b) a C-AEA-731 camshaft, (c) a 997 Cooper exhaust manifold feeding into a Sprite Mk. I silencer and (d) a single 1½in SU mounted on a Janspeed manifold. As can be seen from the power curve, the power output has been more than doubled and since the engine still has exceedingly good bottom end torque, such a conversion is ideal for ordinary road use while power output is high enough for the odd bit of sprinting, or hillclimbing.

Big valve head for fast road or competition use on 850 engines

To do a successful job of this head, as shown in Fig. 65, you need to have a head that can be skimmed by about 0.090in–0.100in so a little sorting will be necessary. We can avoid the necessity of having to select a casting by grinding and drilling into the oil way holes, both the one traversing the head face and the one perpendicular to it, letting in a copper pipe, brazing it in position and resurfacing the head.

Fig 63 Big inlet valve series A head suitable for 850, 997, 1100 and 948 engines

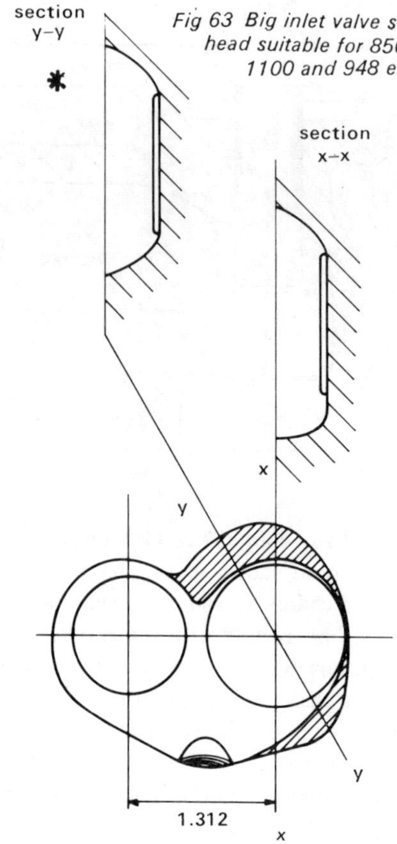

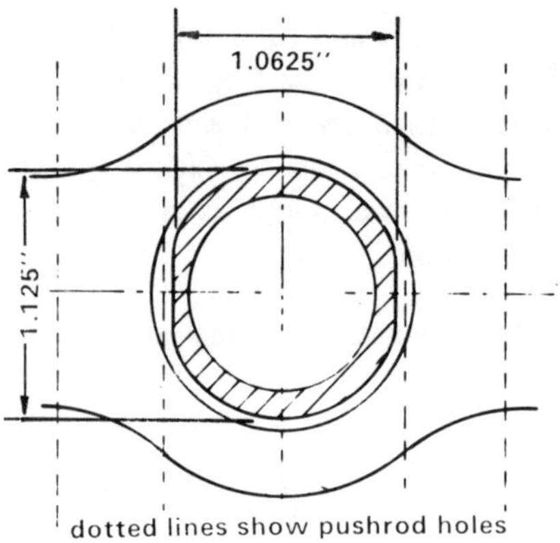

dotted lines show pushrod holes

grind away shaded area to dimensions shown

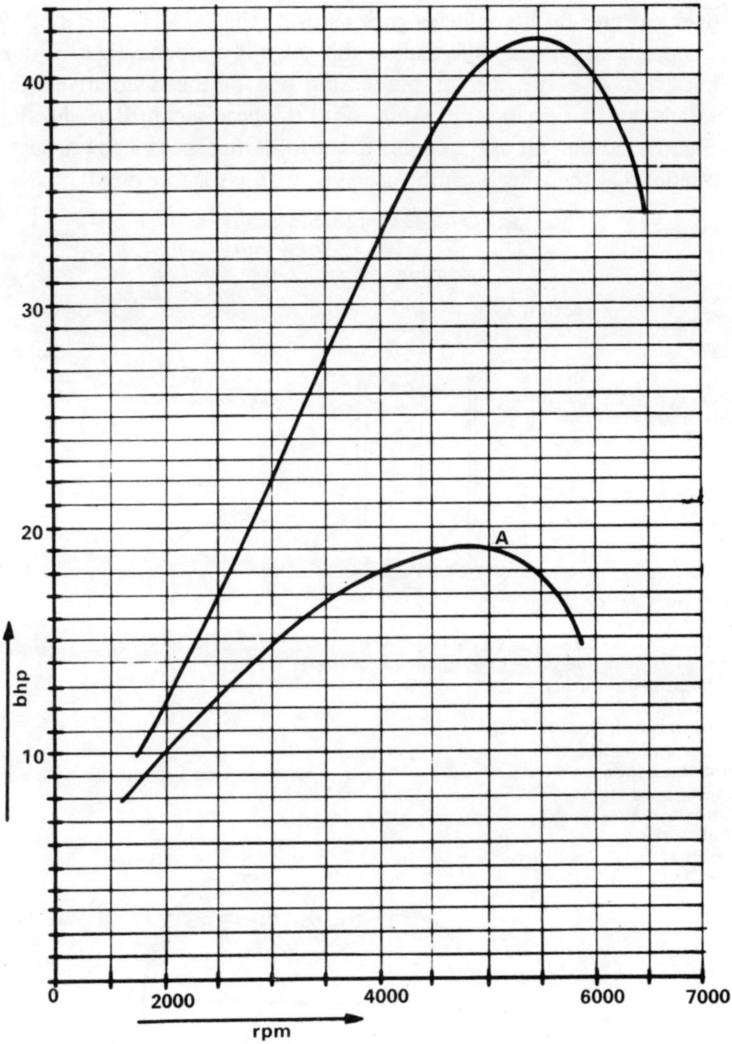

curve A standard output
curve B output using head modified as in fig 63 C–AEA–731 Cam shaft 997 Cooper exhaust manifold with Sprite Mk 1 Silencer—single 1½″ SU. with ram pipe on Janspeed manifold

Fig 64 850 Mini—bhp at the wheels

The use of an 850 Mini head, 997 Cooper or ordinary 1100 head does not give results quite as good as using the MG 1100 head or 998 Cooper head but the difference at this stage is quite marginal. It does, however, save the cost of purchasing one of the alternative heads which can be a financial blessing. With the head modified as shown in Fig. 65 and the rest of the engine to the following spec., a power output of 50bhp at the wheels can be achieved with a silencer fitted.

Fig 65 This chamber shape must be used with 1¼" inlet valves and 1 1/32" exhaust valves

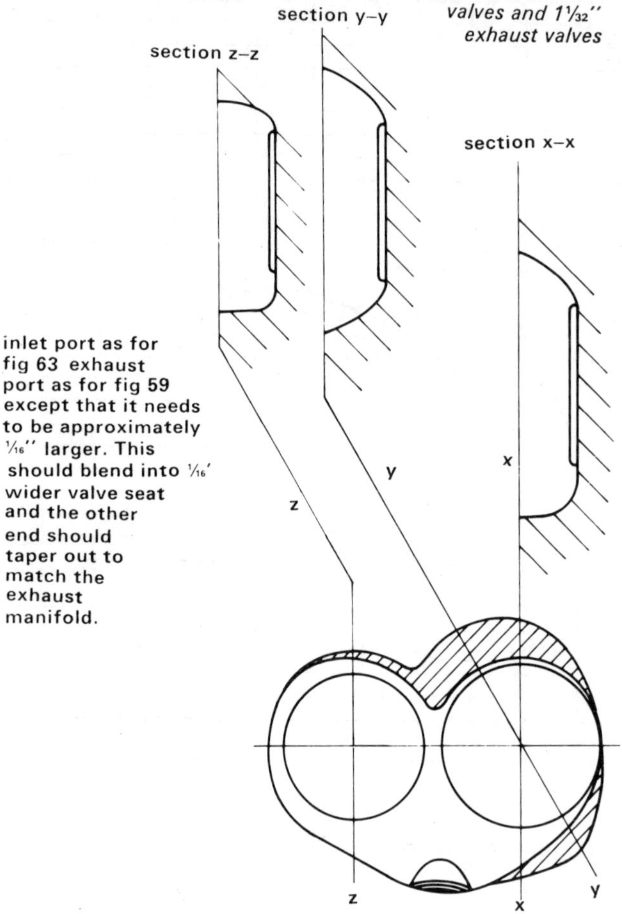

Pistons—flat top solid skirt with 0.0025in extra clearance.

Camshaft—BL competition cam no. C-AEA-648.

Carburation—Weber DCOE 40 on long manifold (as opposed to swan-neck manifold).

Exhaust manifold—Geoff Howe Exhausts. Long centre branch type for 850 Mini.

Silencer—Mk. I Sprite type.

With the silencer removed and the exhaust pipe length adjusted whilst on a dynamometer, the power can be increased to 55bhp. When fitting this head, it is necessary to cut away part of the block as if this is not done the exhaust valve will foul the block. This should be done as shown in Fig. 66.

Fig 66 Cutting away part of the block

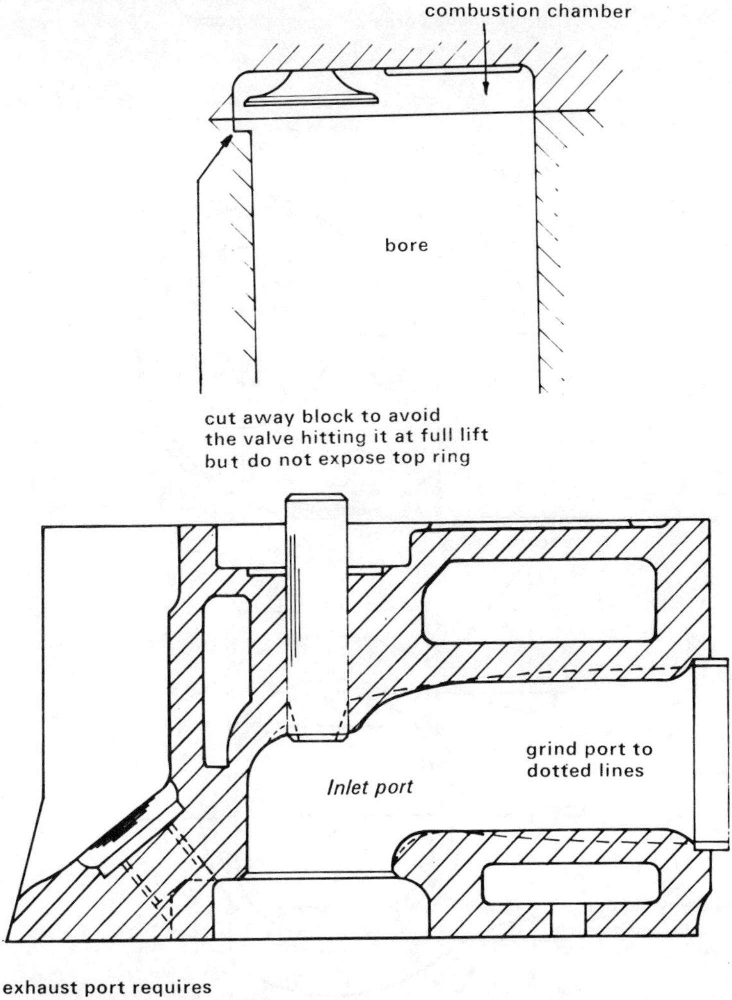

Fig 67 Chamber shape for ✱ 998 Cooper, 1100 MG and Sprites and Midgets from 1964 onwards

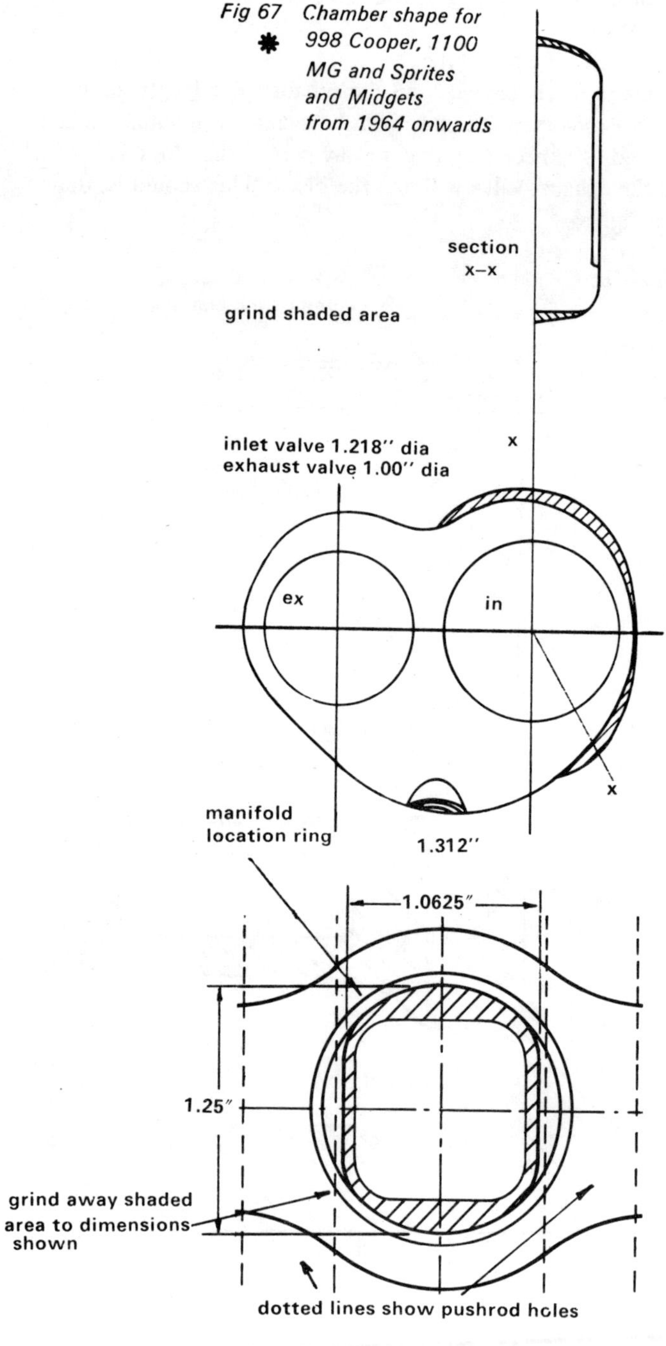

Some readers may be put off by the fact that a full race cam is used with this head, which may in their eyes make it rather unpractical for road use. With a head done to this spec the engine 'comes up on the cam' at 2,800 but with small throttle openings it is possible to use revs as low as 1,500 providing the engine has been set up correctly.

First stage for 998 Coopers, MG 1100's, Sprites and Midgets

The head fitted to 998 Coopers and MG 1100's is basically of a good design for breathing and because of this we are hard pressed to make any really substantial gains whilst still employing the standard valve sizes. The power gains achieved range between 5–7bhp at the wheels depending upon the capacity of the motor to which it is fitted. A lot of the gain in power especially in the mid rev range is due to the increased CR. For road use we should endeavour to get this up around the 10 to 10.5/1 mark. The shapes to modify the head to are shown in Fig. 67.

This head can be used on 850 Minis, 997 Coopers and ordinary 1100's with good results. When used on a Mini it is preferable to make a change of camshaft, carburation and exhaust manifolding. On the 997 Cooper we can use it without any associated modifications but on 1100's some improvement in carburation is advisable.

Full race head for 850, 948, 998 and 1098cc BL engines

To do this head (Fig. 68) we use as a basis an MG 1100, 998 Cooper or late 1100 Sprite or Midget head. To get the best out of this head it should be used together with other competition components such as a full race cam, DCOE 40 or preferably 45 Weber carb, etc.

On 850cc engines using this head, power output just on 60bhp at the wheels has been achieved, while 998 Coopers have shown 69bhp at the wheels and 1100 Sprites and Midget engines 83bhp. It is possible to squeeze out a little more power from all these engines by redesigning the head slightly but in doing so we do cut down on the torque output in certain parts of the rev range especially on the 850 unit, so a little bit of compromising is needed.

As with previously described heads, we need to select a casting that will stand a good deal of skimming, especially when the head is for use on an 850 Mini. For the 850 and 998 engines the valves to use are Cooper S valves turned down to 1.312in inlet and 1.125in exhaust. On 1100cc engines the valves can usefully be increased to 1.360in inlet and 1.156in exhaust. In all cases the block will have to be cut away to give clearance to the exhaust valve .

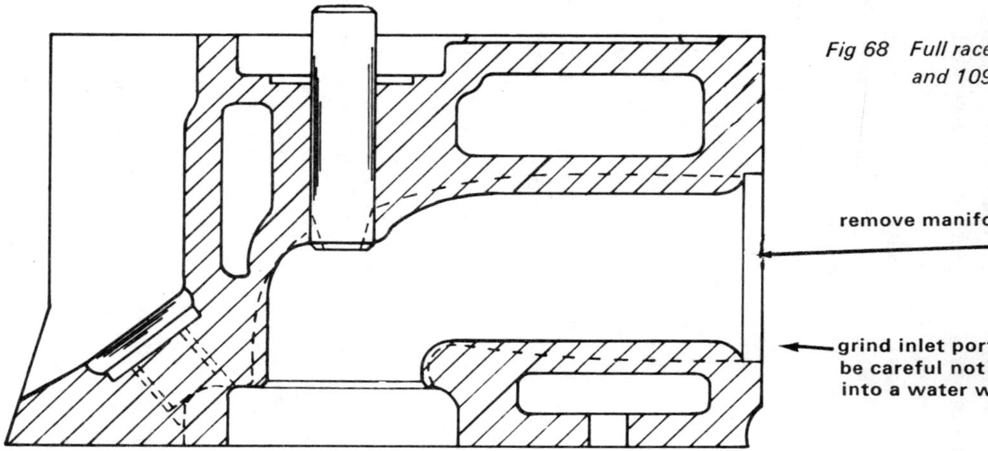

Fig 68 Full race head for 850, 948, 998 and 1098 cc

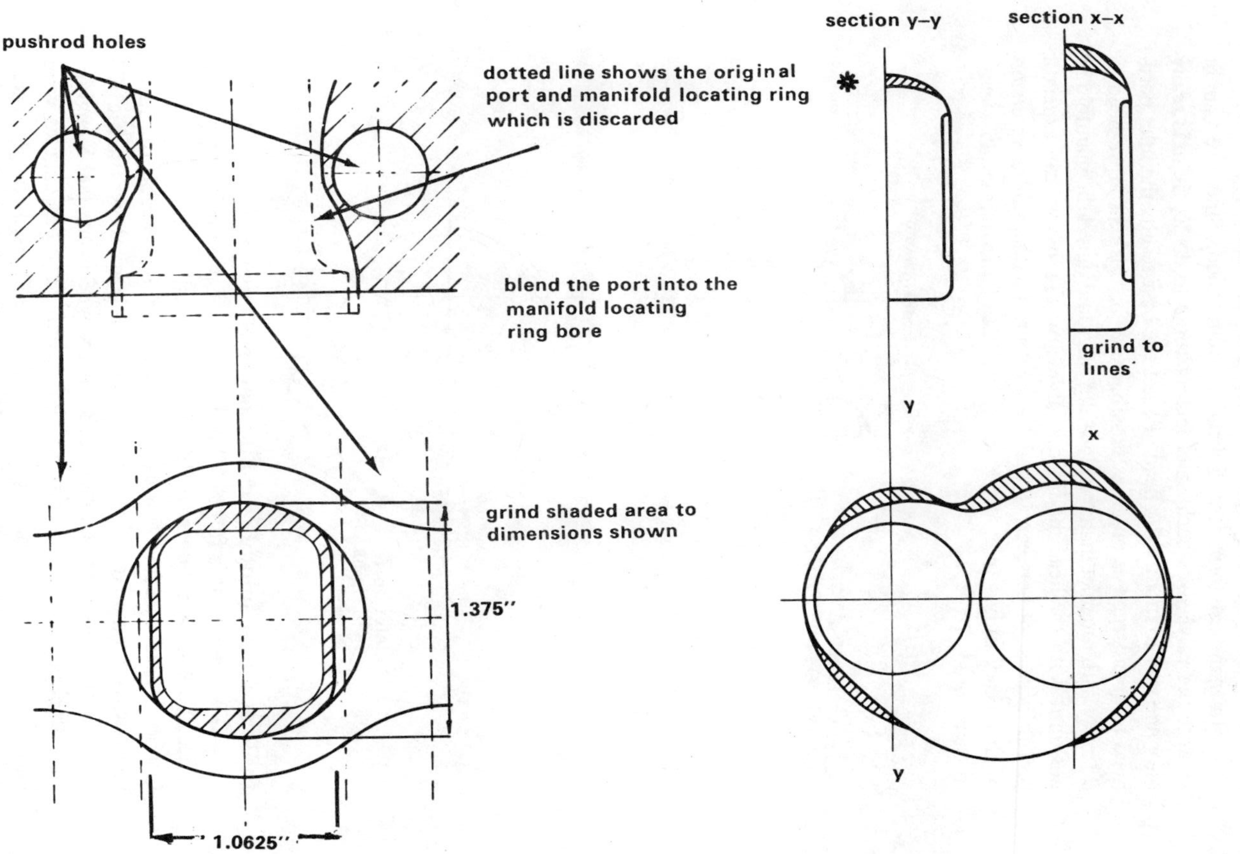
pushrod holes
dotted line shows the original port and manifold locating ring which is discarded
blend the port into the manifold locating ring bore
grind shaded area to dimensions shown
1.375″
1.0625″
section y–y
section x–x
*
grind to lines
y
x
y

Modified Cooper S heads

In standard form, the S head is quite good. Since we cannot enlarge the valve sizes because of lack of room to do so, there is really only one way to modify the head. To get the best results the inlet ports need to be enlarged as much as possible and for this reason we need to do away with the inlet manifold locating rings and dowel the manifold in position. With the head modified as shown in Fig. 69, we can expect a power increase of between 10 and 13 per cent. To do a full race version of this head, radius inlet valve seats, instead of the normal 45° seats, must be used.

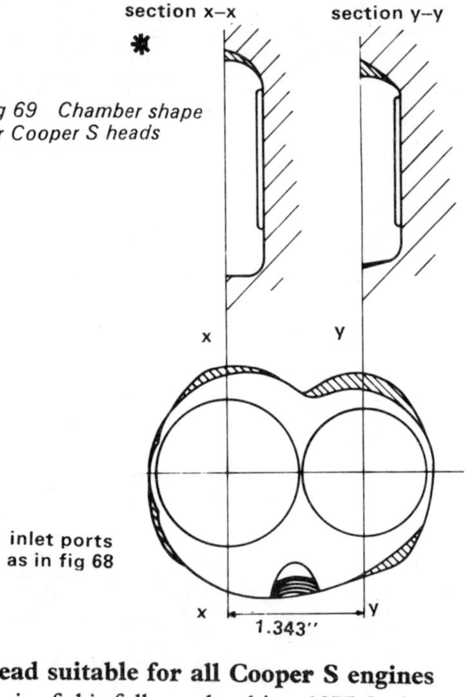

Fig 69 Chamber shape for Cooper S heads

Full race head suitable for all Cooper S engines

The basis of this full race head is a 1275 Sprite or Midget head. Since the standard valves in this head are smaller than those of the Cooper S heads, we find that we have some latitude available to experiment with alternative valve sizes. Using the head shown in Fig. 70 it was found possible to extract 80bhp at the wheels on 1000 S engines and 105bhp at the wheels on the 1300 S unit. The test engines were basically of the following spec: the 1,000cc unit used the BL competition cam C-AEG-595 with a 46 IDA downdraught Weber for carburation, the 1300 unit used the same cam but utilised a 48 IDA Weber. The valve sizes used for this head are 1.406in inlet and

1.156in exhaust. The inlet valve is available under competition part number C-AEG-544 and for the exhaust valve we can use a standard set of 1275 Sprite or Midget exhaust valves.

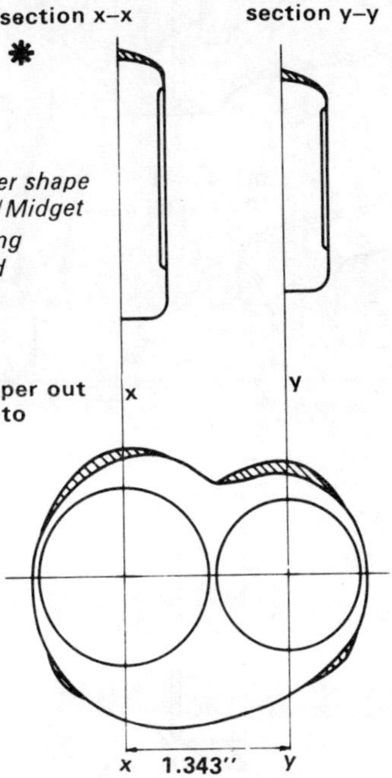

Fig 70 Chamber shape for 1275 Sprite/Midget 1300 heads using 1.406″ inlet and 1.156″ exhaust

FORD HEADS

1000 & 1200 Anglia Heads

The 1,000cc 105E Ford engine has, in its day, been one of the most popular engines for tuning, especially for single seat racing cars. Its modern short stroke design enables it to produce much higher specific power outputs than its contemporary competitors.

The first stage of modification for 1,000 and 1,200 heads is as shown in Fig. 71. To get any immediate worthwhile gains from this head we are presented with a problem. If the head alone is fitted we get a reasonable increase in torque in the mid range revs, but the peak bhp only goes up by about 4. The main reason for this is that the carb is restricting the top end breathing. If a 28/36 Weber is fitted to an otherwise standard engine the power output is increased by about 2–3bhp. If the modified head is fitted to an engine equipped with a 28/36 Weber the resultant power gain is 7–8bhp. The general idea then, is to fit a modified head together with better carburation, thus giving an overall gain of 9–11bhp at the wheels. With a suitable change

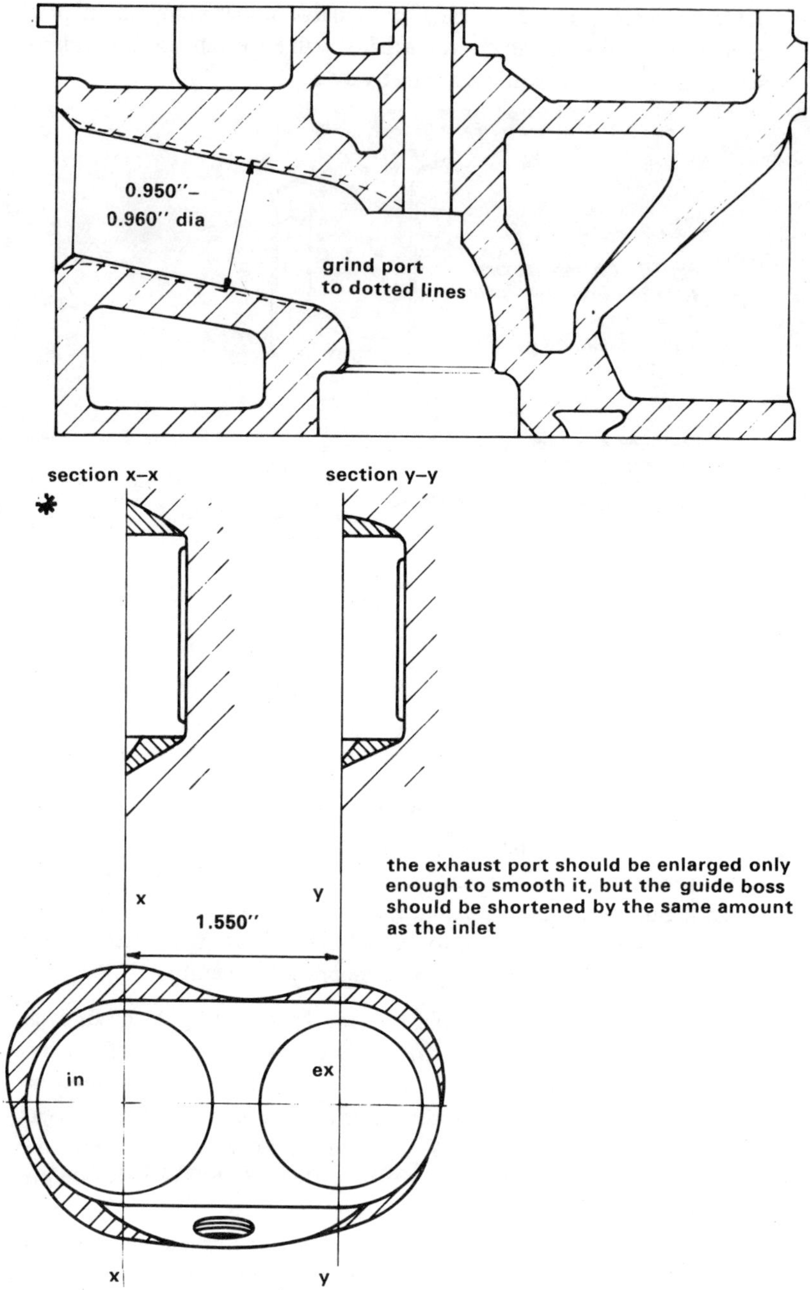

Fig 71 Standard valved 105E combustion chambei

in cam, for instance an A2, WEC or equivalent, a good exhaust manifold and 28/36 Weber, the power output using this head can be pushed to 60–65bhp. If a cam change is made it may be found that coil binding will occur, so steps must be taken to avoid this happening.

Big valve Anglia head

If we are to achieve a substantial improvement in power output from the 1,000cc or 1,200cc engines, a head will be needed that has more valve area. The head design in Fig. 72 fulfils this need. The valve sizes used are as for the 1500 Cortina GT and by using this head together with a pair of 40 DCOE Webers, a Cosworth A6 cam and an extractor exhaust manifold, a power output in the order of 90–95bhp can be expected on a 1 litre engine. For a 1 litre engine quite a bit will have to be skimmed from the head face to get a CR of 11.5/1 or more.

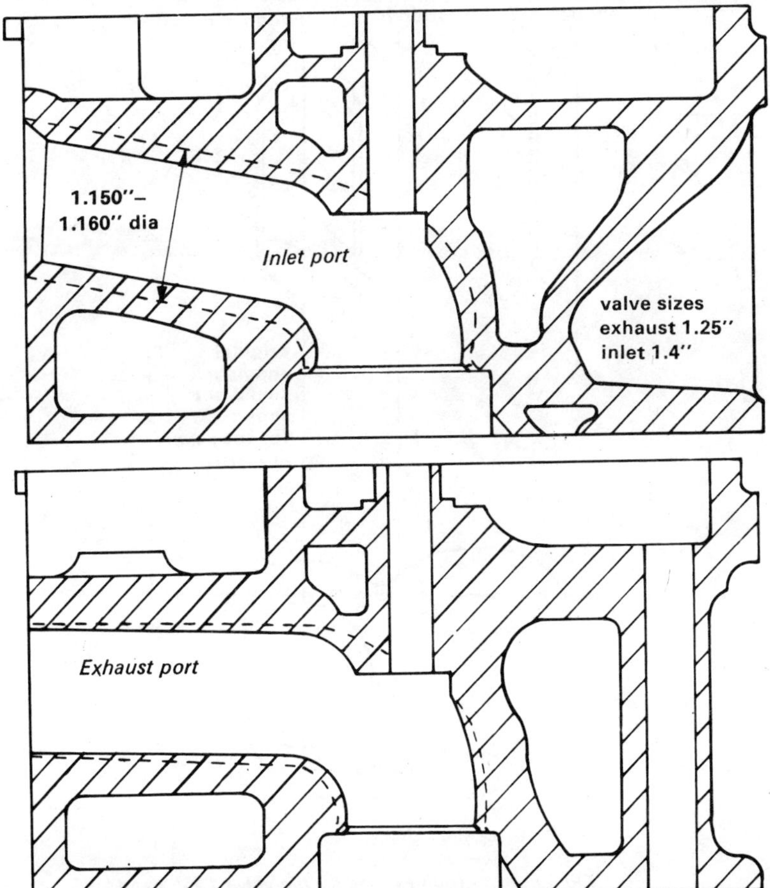

Up to 13/1 can be used. The limiting factor for skimming is the roughly triangular waterway around the spark plug boss. Since the waterway angles back from the head face we can, if the head is skimmed too much, find that the waterway and the sealing edge of the head gasket coincide (Fig. 73). Most heads will stand 0.080in off without any bother but over this figure we should check that the gasket does, in fact, have something to seal on. When selecting a head to use, it's a good idea to make sure that you have not got the low compression head. This has a standard CR of 7.5 instead of 8.5. Apart from using this head as a sort of full race head for 1,000cc and 1,200cc engines, it can, providing the head is not skimmed excessively, be used on 1,340cc and 1,500cc engines. In fact on 1,500cc engines it makes an ideal road-going head especially if the engine is equipped with a 28/36 Weber and a GT cam.

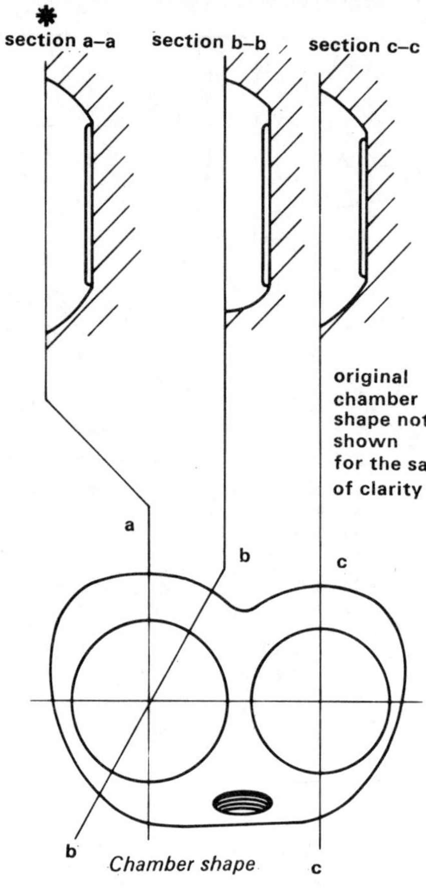

Fig 72 Modifications for big valve 105E head

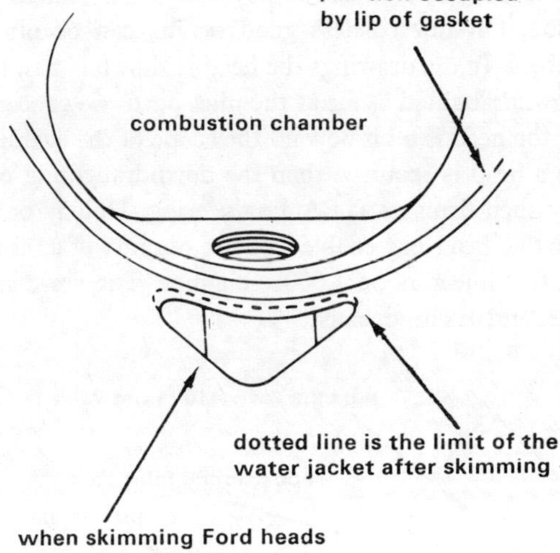

Fig 73 Skimming Ford heads

When this head replaces the ordinary 1500 head we can, on a motor with a GT cam and 28/36 Weber, gain a very full-blooded 16bhp at the wheels. If the head replaces a 1500 GT head, which breathes better than the ordinary 1500 head, we gain 12bhp at the wheels. When used on 1,500cc engines it is essential that the CR be checked to see that it is not too high. For 1,500cc engines, the head must not be skimmed or the CR will definitely be too high. This head is certainly not recommended for 1,650cc engines as it will be found very difficult to get the CR low enough with the increased capacity.

The basic chamber shape can also be applied to the 1,500cc Cortina GT head for use on 1,500cc or 1,650cc engines with almost the same results as achieved by the use of the Anglia head. The basic reason why slightly inferior results are obtained is because the inlet ports are just a shade too big and the chamber shape shown in the drawing cannot quite be achieved, due to metal not being present where it should be. Modifying the ordinary 1,500cc non GT head is not really worth the trouble. The best idea, if this situation presents itself, is to swop it for a 997cc head and modify that.

Full race head for 1,000cc or 1100cc (85mm bore) 105E engine

The head design depicted in Fig. 74 is intended for use on engines where the maximum power is sought. It is ideally suited for use with fuel injection, but nevertheless good results can be obtained on 40 DCOE Webers. In the drawings the head is shown in two forms, one of which is downdraughted as far as the inlet ports are concerned. Downdraughting the head is a bit beyond the scope of the average enthusiast but if such a head is required then the downdraughting operation can be done by such firms as WRA Engineering, Holbay or Broadspeed. The use of this head has enabled power outputs of 114bhp on carbs, 120bhp on fuel injectors on 1,000cc engines. The valve sizes used are 1.410in inlet and 1.25in exhaust.

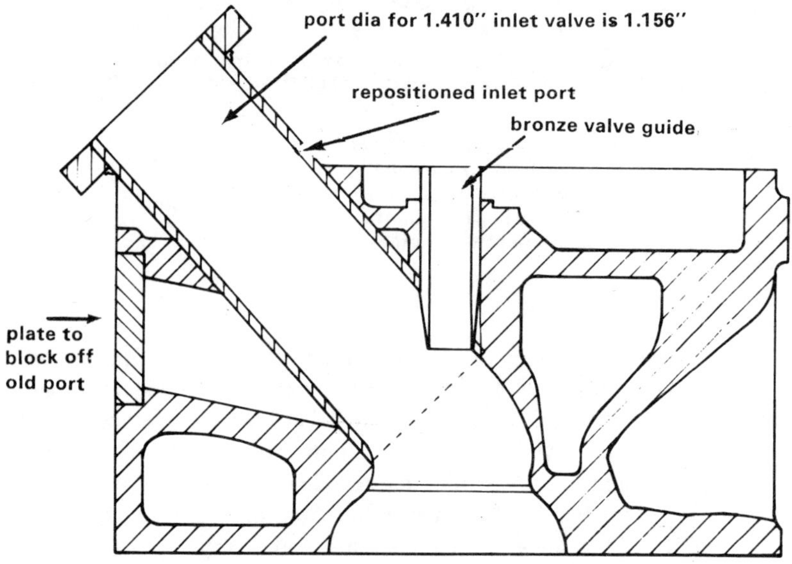

Fig 74 Down draughted 105E head

Full race head for 1,300cc plus engines

As with the previous head we have the situation where this head can either be downdraughted or not (Fig. 75), depending on the reader's finances. The basis of this head for 1,300cc and 1,500cc engines is a 105E head, but for engines of larger capacity we should use a 1500 Cortina GT head. On 1,300cc engines the head will accept a CR of 13.2/1. On 1,500cc engines this should have the CR around 12.8/1.

Fig 75
Chamber shape A
for 1500 GT head

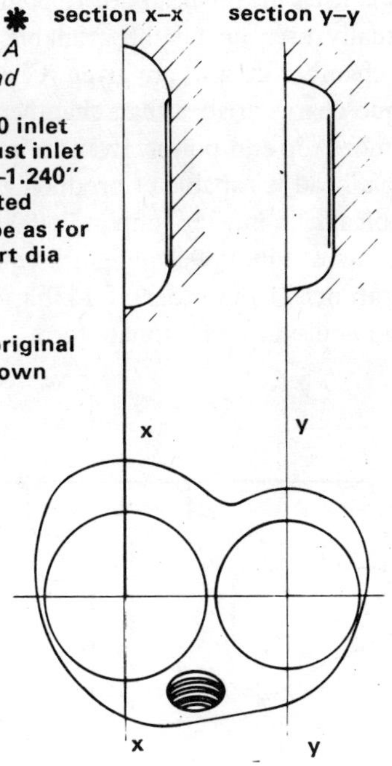

valve sizes 1.560 inlet and 1.350 exhaust inlet port dia 1.220"–1.240" a down draughted inlet port will be as for size exhaust port dia will be 1.125"

for clarity the original shape is not shown

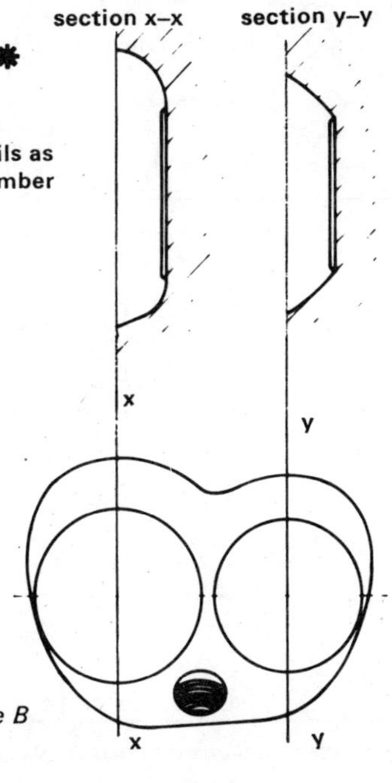

all other details as for fig 75 chamber shape A

Fig 75
Chamber shape B

The valve sizes used in this head are 1.560in inlet and 1.350in exhaust and this virtually uses up all the available space for valves. In this instance two chamber designs are given. Chamber shape A has slightly superior torque characteristics than chamber B but shape B does produce a little more top end power. Even in its basic form, ie not downdraughted, this head is capable of producing high power outputs. Just to quote a typical example, 161bhp at 7,400 revs was extracted from a 1650 unit on carbs. On 1,300cc engines equipped with fuel injection etc., power outputs in the order of 145bhp at 9,000rpm are possible when the head is used in its ultimate form.

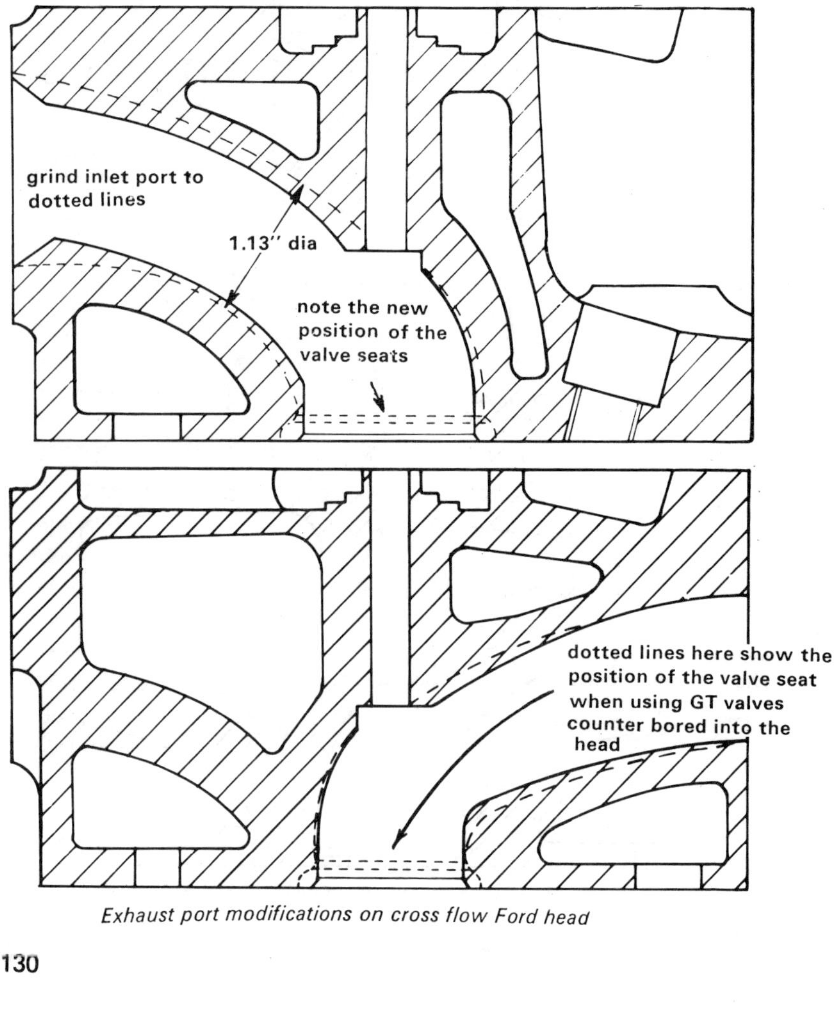

Exhaust port modifications on cross flow Ford head

Crossflow Ford heads

With the exception of the small combustion chamber recesses on the GT heads, all crossflow heads are basically the same. The first stage of modification to these heads is to gas flow the inlet and exhaust ports. Since these heads are quite good in standard form we find, by flowing the ports, only a small increase in power, usually about 5%

Fig 76 1100 and 1300 non GT head modified to take GT valves
Combustion chamber for cross-flow Ford

NB no chamber mods necessary if GT head or 1100 or 1300 heads with standard size valve

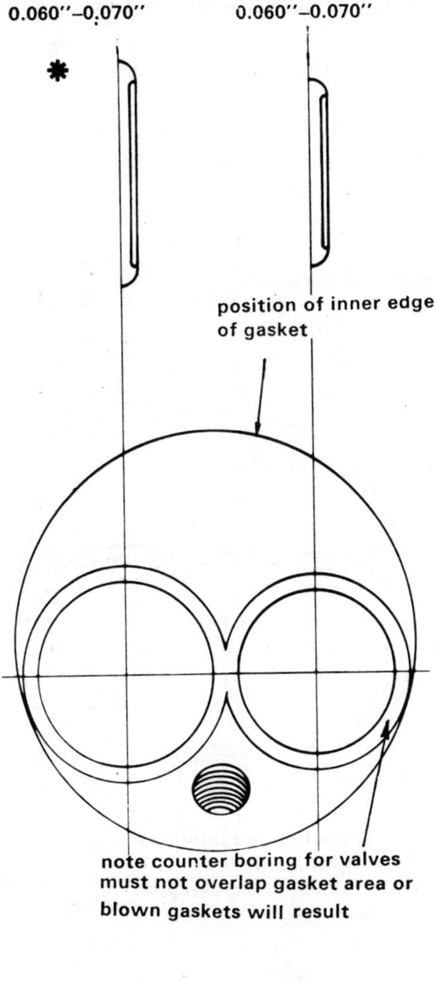

The main difficulty with these heads is that it is not a simple straightforward skimming job to achieve a higher CR. On the non GT head there is no chamber whatsoever in the head, therefore no amount of skimming will give any increase at all in the CR. On the GT engine we do have a very shallow chamber, but if the head is skimmed to any degree worth mentioning, the valves will strike the pistons. Any increase in CR on non GT engines will have to be achieved by a change in pistons, but we can, by modifying a non GT head for use on a GT engine, achieve a small increase in CR.

By using an 1,100cc or 1,300cc head as a basis, we can modify the head (Fig. 76) to accept the larger valves of the GT head which, incidentally, are also shorter than the non GT valves. By modifying the head as in Fig. 76 we gain power in two ways. First, the CR is increased slightly, and secondly a superior gas flow is achieved by countersinking the valves into the head when the correct profile is used. Strictly speaking, the head shown in Fig. 76 is also best on non GT engines but for the 1,100cc engine it is a bit over valved unless other mods are done, such as fitting a hotter cam, better carburation and free flow exhaust. On the standard 1,300cc it is advisable to use at least a GT cam and Weber carb.

Modifying Imp heads

The best basis to use for a Stage 1 head upwards as far as Imps are concerned is the big inlet valve Imp sport head. We can just use the earlier small inlet valve head and have the larger valves fitted, but it is not entirely satisfactory if the engine is tuned to a fair degree because it is slightly lacking in stiffness. Assuming we are starting with a 1.28in inlet dia. head, it should be modified as shown in Fig. 77.

On an 875 or 1000 Imp with twin SU's or Strombergs, this will show an increase of 6–7bhp depending upon capacity. When raising the CR on Imp engines, we must keep an eye on the piston to valve clearance especially if a cam change has been made. Most heads can be skimmed 60–70 thou. and this usually raises the CR to about 11/1. With CR above this figure, we can get head distortion, when under load, resulting in blown head gaskets. The solution is to fit larger diameter head bolts and machine a 0.020in wide by 0.010in deep groove around the top of the cylinder liners such that the groove coincides with the centre of the lip around the bores of the head gasket.

If a head for competition purposes is required and the ultimate is sought whilst still using an Imp sport head, then the head should be modified as shown in Fig. 78. The first operation on this head is to remove the valve inserts and weld metal into the chamber so as to

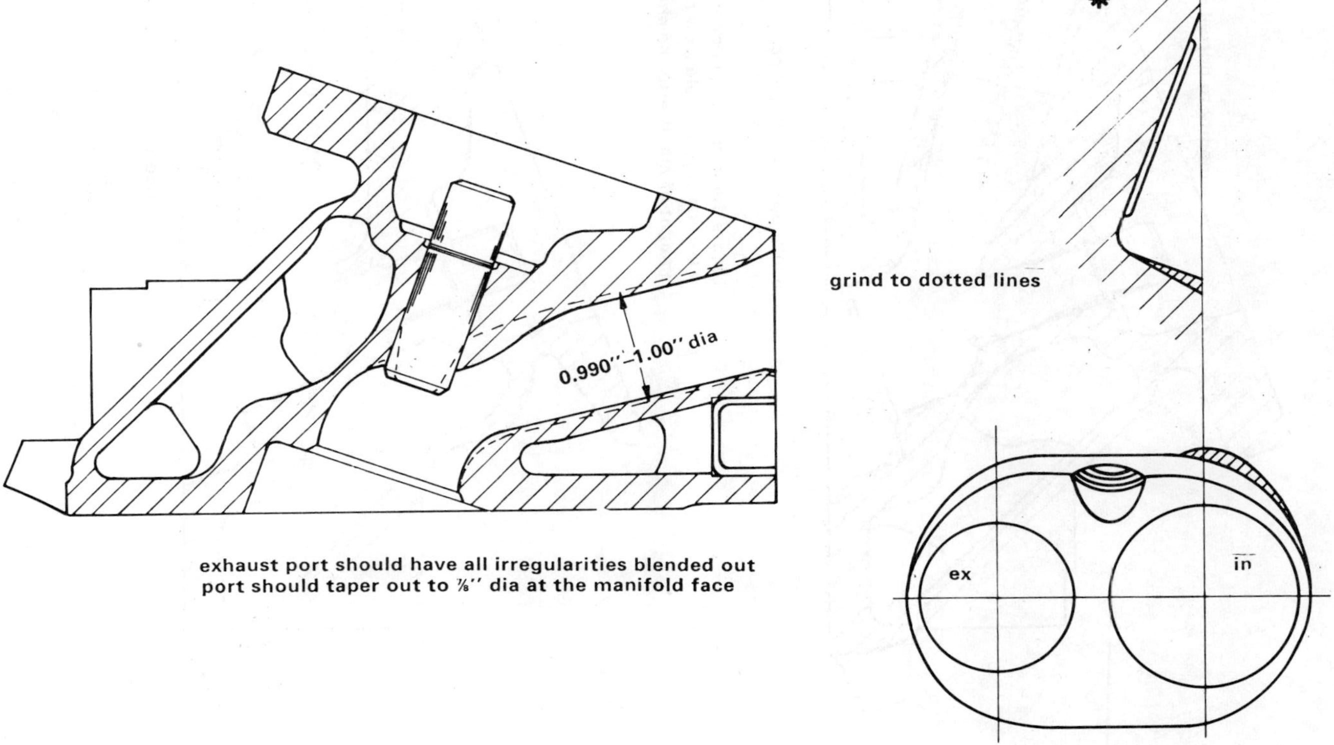

Fig 77 Standard valved Imp Sport inlet port modification. Modification to Imp Sport Chamber

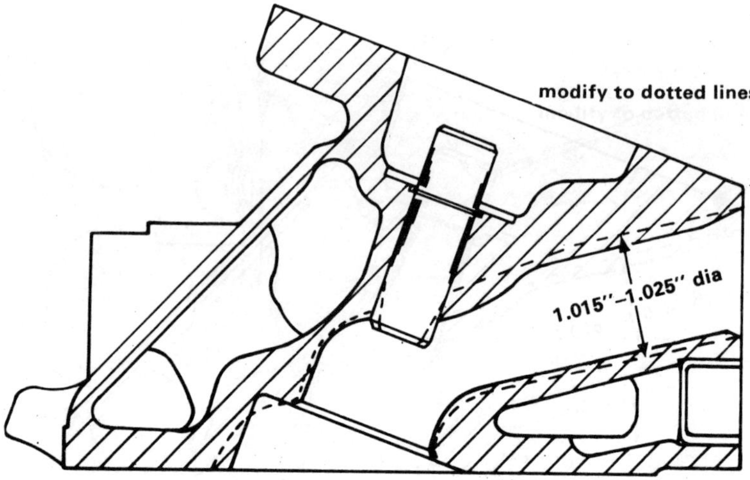

Imp inlet port for full race head

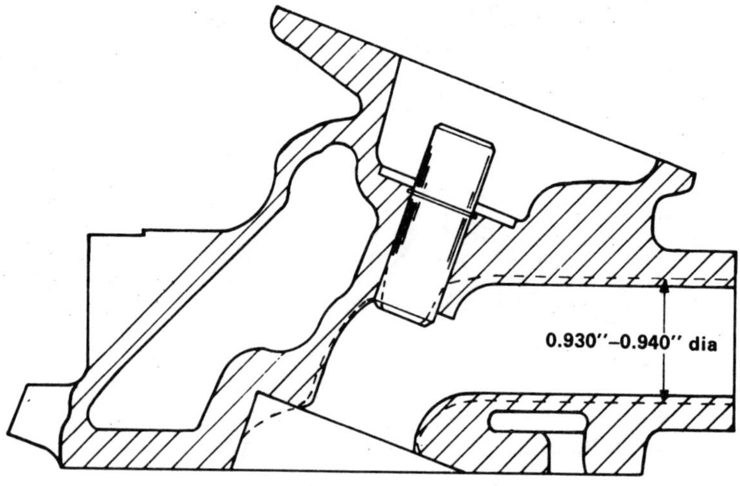

Imp exhaust port for full race head

Fig 78 Imp chamber shape for full race engine

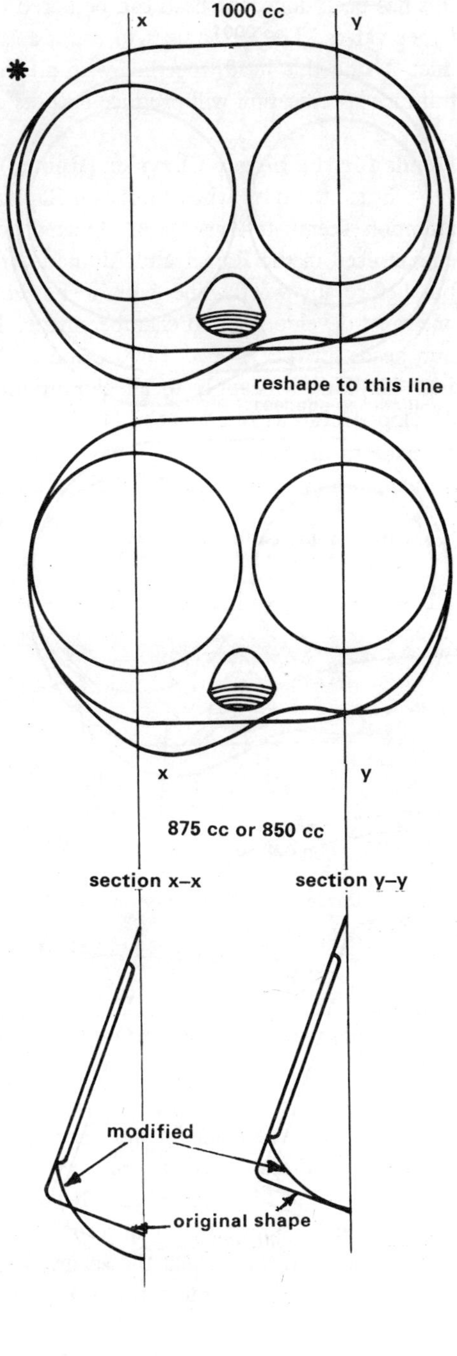

allow the curved section of the chamber wall to be produced. Once this has been done, the head can be bored to accept the inserts for the larger valves. The valve sizes to use are 1.125in exhaust and 1.312in inlet. Using this head, together with all the rest of the engine up to full race specification will produce outputs of 114bhp.

Heads for the bigger Chrysler (Rootes) engines

Since the days when Sunbeam Rapiers were top rally cars, little attention seems to have been directed towards tuning them. The engine used in the Rapier and Alpine is, in fact, quite a good motor, but before any worthwhile gains in power can be achieved, it seems one must develop a sound chamber shape. The way to modify aluminium heads for the 1,500cc, 1,600cc and 1,725cc engines is as shown in Fig. 79. This will apply to all aluminium heads except the Holbay developed head as fitted to the H120 Rapier. Little improvement can be made to the Holbay head without extensive work, so the best plan is to smooth out ports and chambers and leave it at that.

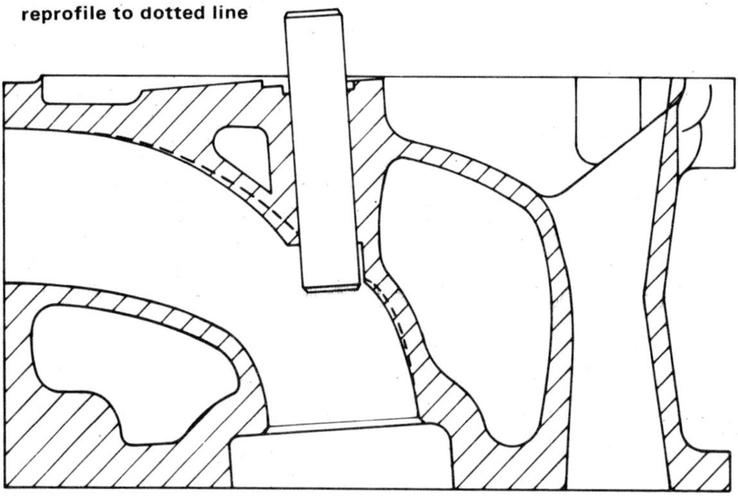

Modified inlet port for Chrysler 1500, 1600 and 1725 alloy heads

The cast iron heads of the larger Chrysler (Rootes) engines are of basically differing designs from the alloy ones. These should be modified as shown in Figs. 80 and 81. The limiting factor for all these heads as far as skimming is concerned is the tapped holes along the base of the head into which the securing screws for the tappet chest cover pass. When skimming the head, we must avoid breaking into

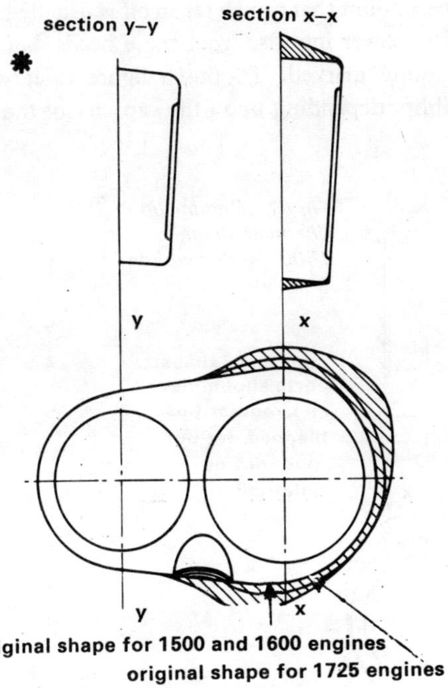

Fig 79 Modified chamber shape for Chrysler 1500, 1600 and 1725 alloy heads

Fig 80 Chamber shape for 1390 Minx head

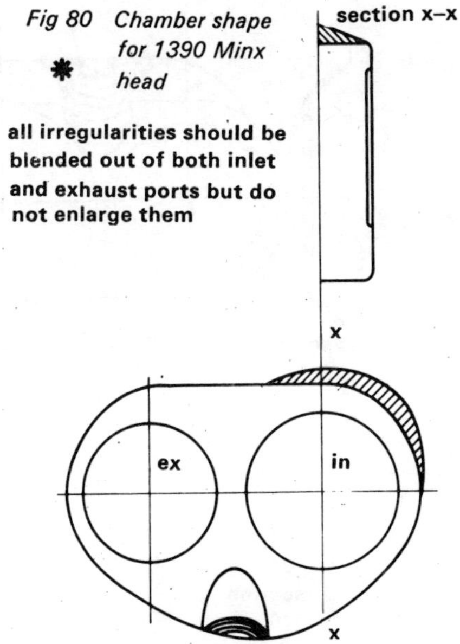

these holes. The amount that can be taken off is usually between 0.060in and 0.075in. The power increase from these heads, both cast iron and aluminium, is quite marked. To put a figure to it we can achieve between 11–15bhp, depending upon the capacity of the engine.

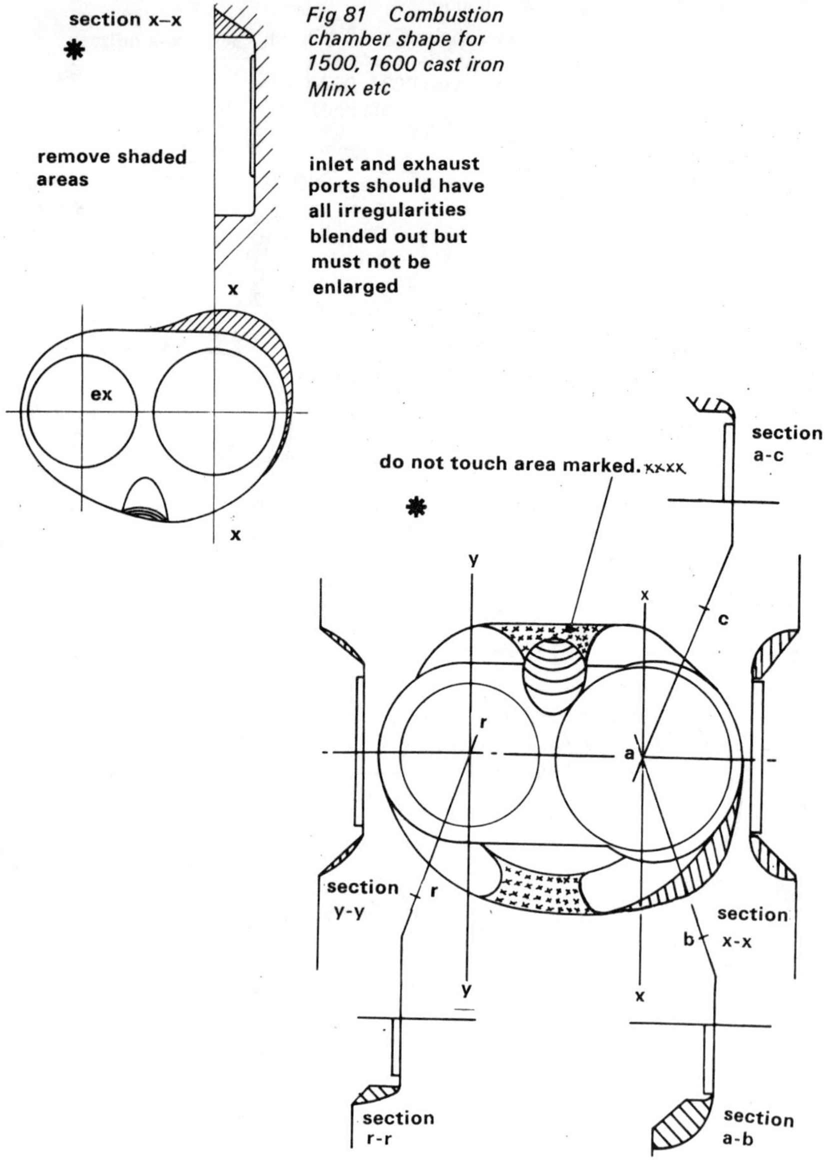

Fig 81 Combustion chamber shape for 1500, 1600 cast iron Minx etc

Triumph heads 1600 and early 2 litre engines

The 1,600cc and early 2,000cc engines, as fitted to Mk. 1 Triumph 2000, 1600 Vitesse and Mk. 1 2 litre Vitesse, have a different head casting from the Mk. 2 versions and 2.5 litre. The modifications to the Mk. 1 type of head is as shown in Fig. 82. When modifying this head for use with standard size valves, we must not enlarge the inlet ports any more than is necessary to achieve a smooth profile as they are too big as standard. Once the head has been reshaped and polished it can be skimmed by up to 0.080in to raise the CR. A head modified as per

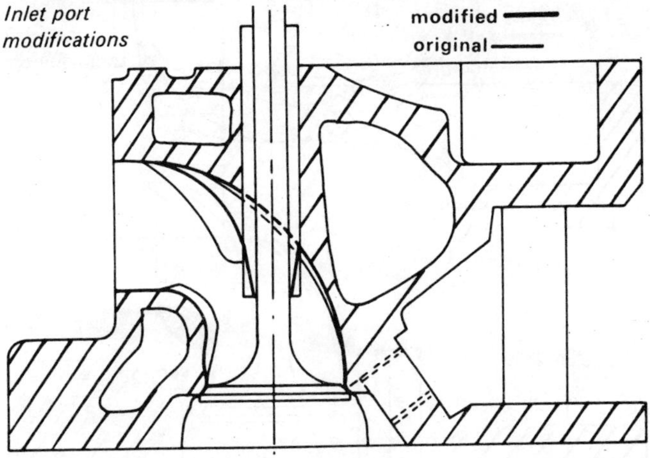

Fig 82 Modification to 2000 Triumph combustion chambers stage 1

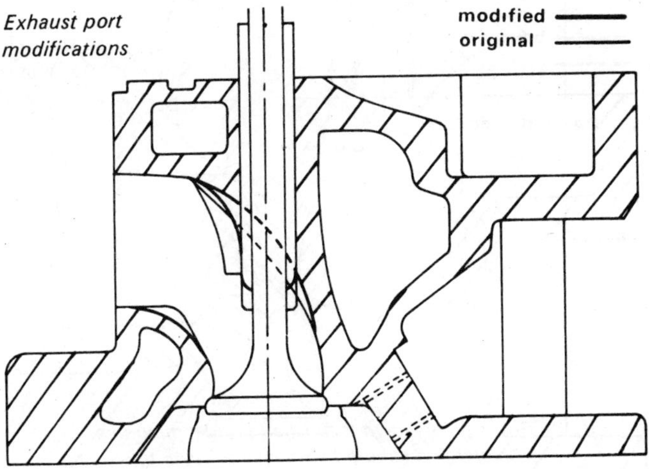

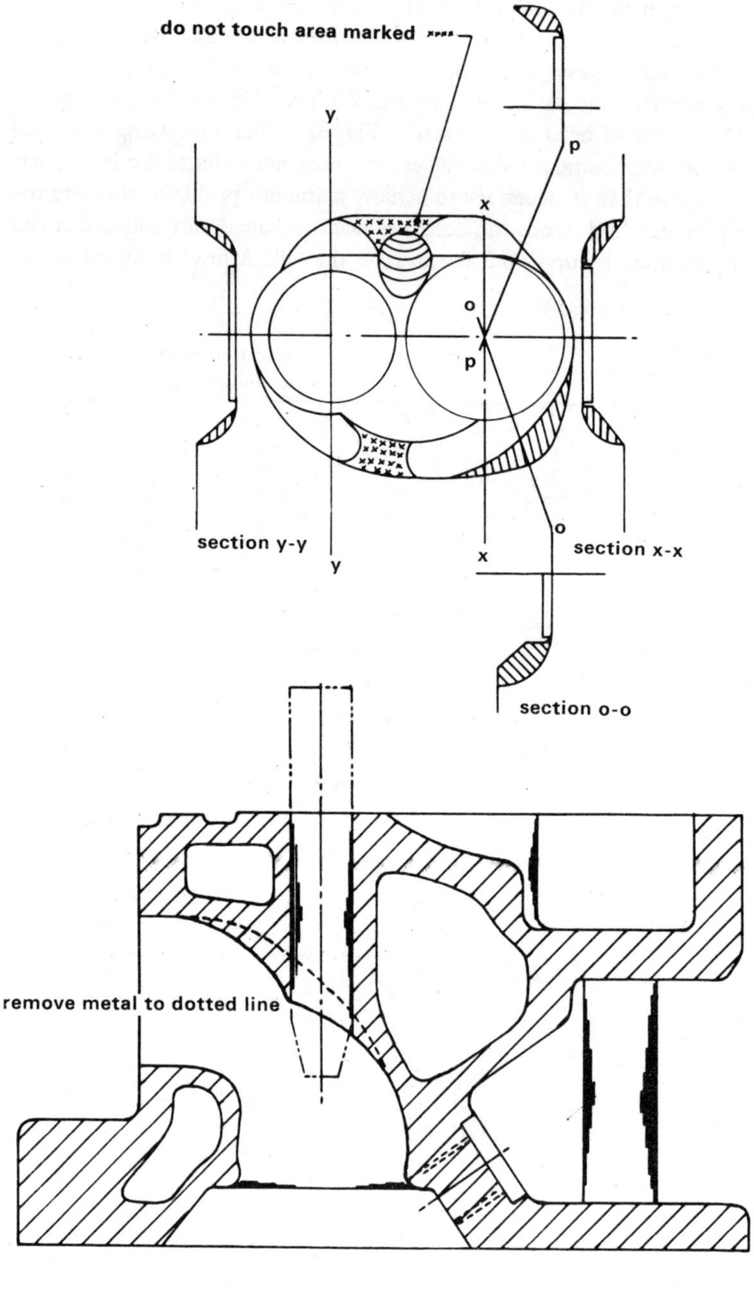

Fig 83 Port shape shown is finished profile, minimum enlargement should be used to achieve this shape

drawing will give between 4 and 8bhp depending upon the type of carbs fitted. If the engine is going to be modified more drastically by fitting alternative cams, exhaust manifolds, etc., then we can beneficially use a larger inlet valve. The exhaust port mods remain as for the standard inlet valve head shown in Fig. 82 but the chamber shape and inlet port shape are slightly different, and should conform to Fig. 83, Even with the larger inlet valve, 1.395in instead of the standard 1.3in. we should not enlarge the first $\frac{1}{2}$in of the port any more than is necessary to match it up to the manifold.

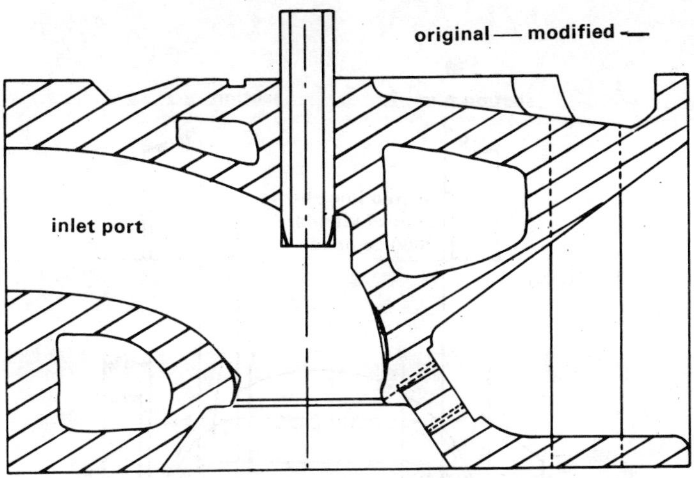

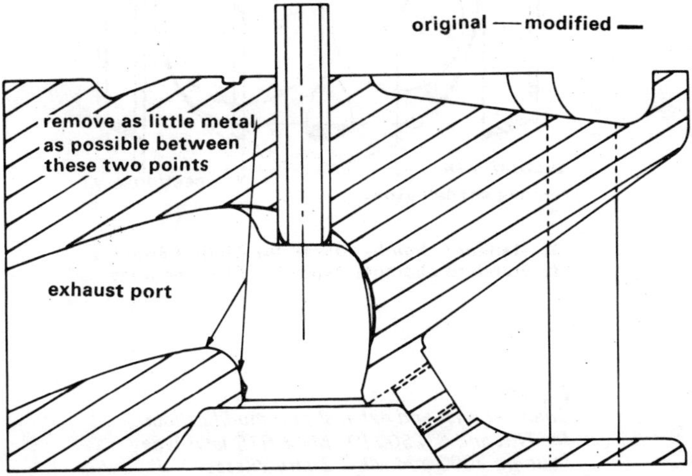

Mk. 2 2-litre and 2.5-litre engines

The head fitted to the later Triumph engines is of a generally superior design. The inlet ports have a much more gentle curve as they approach the inlet valve and the inlet valve itself is over $\frac{1}{8}$in larger. The combustion chamber has also been redesigned and as such there is very little we can do to improve it. The head should be modified as shown in Fig. 84. By doing to, we can expect a power increase of about 7bhp on the 2-litre engines and 9bhp on the 2.5-litre engines.

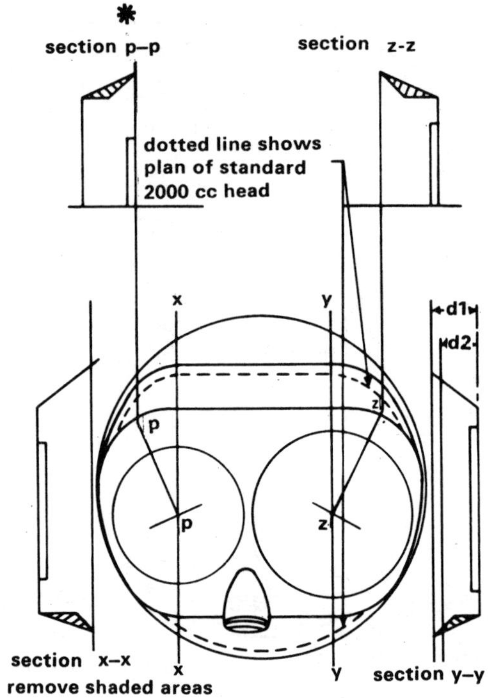

d1 standard chamber depth for 2.5 litre engines
d2 standard chamber depth for 2 litre engines

Fig 84
Chamber inlet and exhaust port modifications for TR5 and 6, 2500 P1, Mk 2 GT6 later type Triumph 2000 and Mk 2 2-litre Vitesses

Triumph TR3, 4 & 4A engines

Modification to the large four cylinder Triumph heads is relatively straightforward. When modified as shown in Fig. 85 we can expect a power increase of 10–12bhp depending upon capacity. To raise the CR, the head can safely be skimmed by 0.080in. The advantages of this head make it eminently suitable for road work when it is the only mod. to the engine. Both mid range torque and flexibility are improved, thus allowing faster hill climbing and easier town driving. This head is also suitable for use with other mods. such as a pair of 40 or 45 DCOE Webers, full race cam, etc. In full race trim with this head, we can expect a power output of 142–145bhp from a 2-litre version and over 150bhp from a 2.2-litre unit.

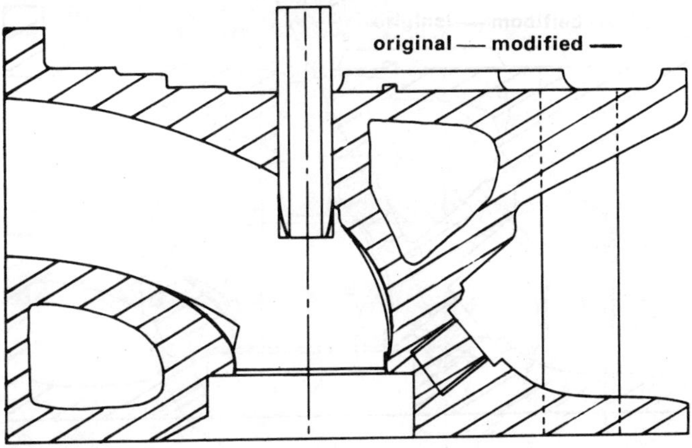

Inlet port

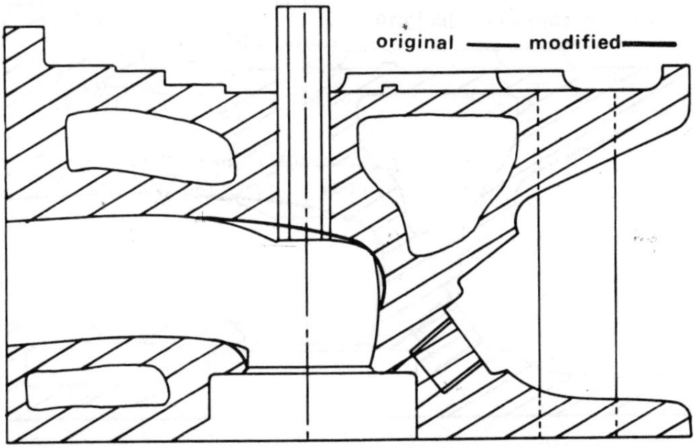

Exhaust port

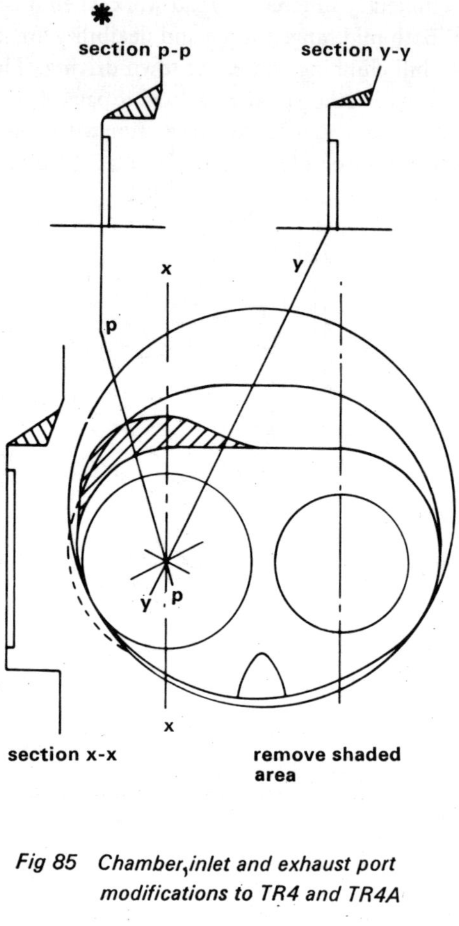

Fig 85 Chamber, inlet and exhaust port modifications to TR4 and TR4A

Appendices

appendix A

Spark Plug Specifications

D Series ½″ reach 18mm thread

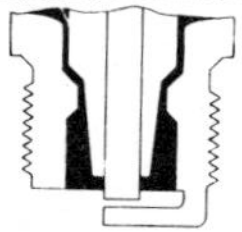

HOTTER	AC	Autolite	Bosch	Lodge	NGK	KLG
D-9	C82					
	M83	BT3	M225T1	HIP	A-7	M80
D-6	C81	BT2	M240T1	H318	A-7	M100

F Series conical seat (no gasket) 18mm thread
(Do not overtorque. Install at 15-18 ft./lbs.)

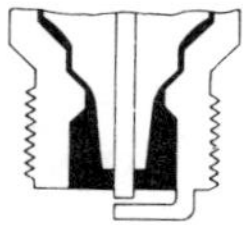

HOTTER	AC	Autolite	Bosch	Lodge	NGK	KLG
F-10	C83T	BTF3				
	C84T	BF32	MA175T1	HTN	A-7F	TMT50
		BTF31				
F-82		BTF1		HTN18		
		BR22				

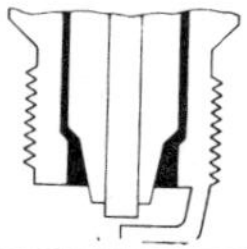

	AC	Autolite	Bosch	Lodge	NGK	KLG
F-9Y	C83T	BF32		HTNY		
F83Y		BF22			A-7F	

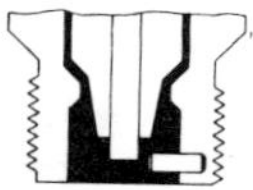

	AC	Autolite	Bosch	Lodge	NGK	KLG
F-62R		BF703				
F-60R		BF601				
F-57R		BF603				
F-54R		BF403				
		BF203				

G Series .700″ reach 10mm thread

HOTTER		AC	Autolite	Bosch	Lodge	NGK	Marchal
G-63 G-61	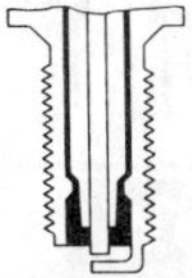						846 847
G-59R G-56R G-54R	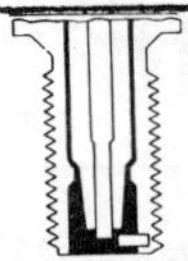		PG603 PG403	U270T17 U310T17 U400T17	10RL49 10RL50 10RL51		847

J Series $\frac{3}{8}$″ reach 14mm thread

HOTTER		AC	Autolite	Bosch	Lodge	NGK	KLG
HO-8A or UJ-11G HO-3 or UJ-7G UJ-64P UJ-60P	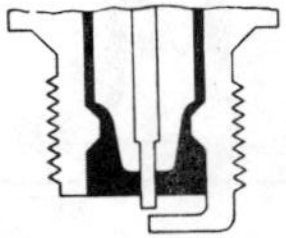		AZ2				
J-10Y J-63Y J-61Y or J-86Y† J-83Y†	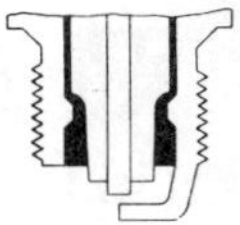	C44S 42S 43S C421 C42	AT42 A32 A22 A12	W145T7	CANY	BP-6	FS-55P
J-6J J-4J J-2J	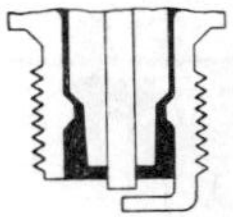	M44C 43 M43 C43 42 C42-1 M42K M41K M41G	AT3 A3X A3 A3XM AT1	W225T3 W240T3 W260T3	HAN HANP 2HAN 3HAN	B-77C B-7 B-76	FS-75H FS-100H FS-100

Plug		AC	Autolite	Bosch	Lodge	NGK	KLG
J-62R J-60R			A23 A903	W240T16 W270T16	R-47	B-8N B-9N	
K Series ½″ reach 18mm thread							
HOTTER K-60R K-57R K-54R	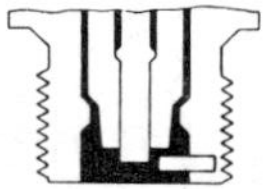	AC M82	Autolite B903 B603 B403 B203	Bosch M270T16	Lodge	NGK AB-83K AB-93K	KLG M83 M82 M81
K Series .445″ reach 18mm thread							
UK-10 K-9 UK-7	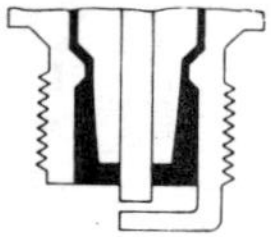	M84 M83	BZ3	M175T1 M225T1	HTN18	AB-6 A-7	
L Series ½″ reach 14mm thread							
K Series .625″ reach 18mm thread							
K-83R K-80R K-78T	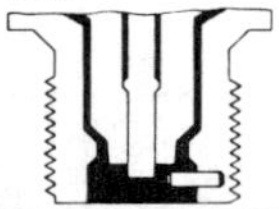		BN603 BN403 BN203				
HOTTER L-62R L-60R	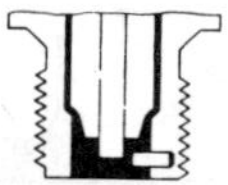	AC	Autolite AE23 AE903	Bosch W175T1 W225T1	Lodge R-47 R-49	NGK B8HN B9HN	KLG F-220 F-250 F-260

L Series ½″ reach Special types methanol nitro Outboards and Racing-carts

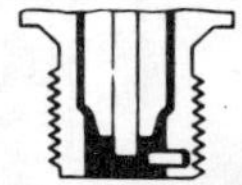

	AC	Autolite	Bosch	Lodge	NGK	KLG
L-87R	M-40					
	17-41FF					
L-84R						
L-83R						

L Series ½″ reach 14mm thread

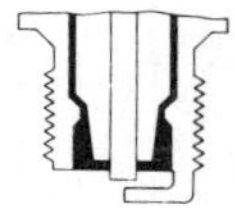

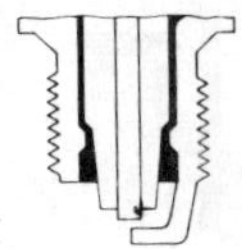

HOTTER	AC	Autolite	Bosch	Lodge	NGK	KLG
L-7		AE4		HF2H		
	44F	AE3	W175T1	HN H14	B6H	F-75
L-5			W225T1			
	42F	AE2	W240T1	2HN	B7H	F-100
L-4J		AE2	W175T1	2HN	B7H	F-100
L-85		AE4				
	46FF	AE3	W225T1	HF2H	B6HZ	
L-81	M42F		W240P11S	HH14	B7H	DF-80
UL-12Y	B44FS	AE42			BP6H	
L-66Y		AE32				
L-64Y		AE22				
L-61Y						
L-87Y	45F					
	45FD	AE32	W175T7	CNY		F-55P
L-82Y or UL-82Y	42F	AE22	W225T7			

N Series ¾″ reach 14mm thread

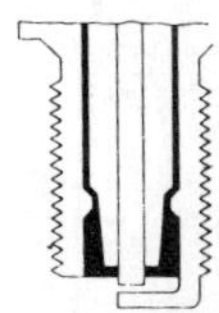

HOTTER	AC	Autolite	Bosch	Lodge	NGK	KLG
N-4	C42N					
	C41N	AG2	W225T2	2HLN	B-7E	FE80
N-3				HLN		
	43XL	AG901	W240T2	2HLN	B-7E	FE100
				HF42L		
N-60						
N-57						

N-9Y N-66Y N-65Y N-64Y N-63Y N-60Y	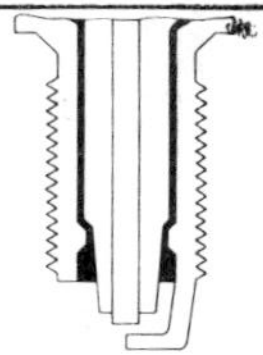	C42N	AG32 AG22 AG12	W200T30	HLNY 2HLNY 2HLNY 3HLNY 4HLNY 5HLNY	BP7E	FE65P F125P
N-62R N-60R N-57R N-54R N-52R	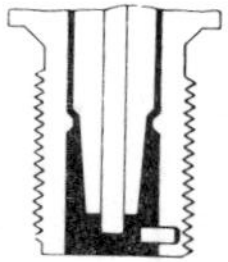		AG23 AG903 AG603 AG403 AG103 AG203	W240T17 W270T17 W310T17 W400T17 W440T17	RL47 RL47 RL49 RL51 RL53	B-8EN B-8EN B-9EN B-10EN B-12EN	FE220 FE260 FE280 FE310 FE340

R Series ¾″ reach 12mm thread

HOTTER		AC	Autolite	Bosch	Lodge	NGK	Marchal
R-63 R-61	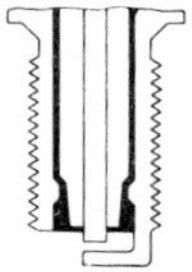			 X240T17		 D8E	 810
R-59R R-56R R-54R R-52R	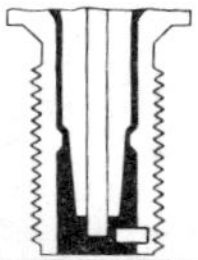			X270T17 X310T17 X400T17 X440T17		D9EN D10EN D12EN D13EN	810 804

FACTORY GAP TOLERANCES

D	
D-6	.023-.028″
D-9	.023-.028″

F	
F-9Y	.033-.038″
F-10Y	.028-.033″
F-11Y	.028-.033″
F-54R	.013-.018″
F-57R	.018-.023″
F-60R	.018-.023″
F-60Y	.018-.023″
F-62R	.018-.023″
F-62Y	.018-.023″
F-79Y	.018-.023″
F81Y	.018-.023″
F-82	.023-.028″
F-83Y	.028-.033″

G	
G-54R	.011-.014″
G-56R	.011-.014″
G-59R	.011-.014″
G-61	.011-.014″
G-63	.011-.014″

J	
J-2J	.023-.028″
HO-3	.023-.028″
J-4J	.028-.033″
J-6J	.028-.033″
J-7J	.023-.028″
UJ-7P	.023-.028″
UJ-10Y	.028-.033″
UJ-11P	.023-.028″
J-54R	.013-.018″
J-57R	.018-.023″
UJ-60P	.018-.023″
J-60R	.018-.023″
J-61Y	.018-.023″
J-62R	.018-.023″
J-63Y	.018-.023″
UJ-64P	.023-.028″
J-79	.018-.023″
J-86	.018-.023″

K	
UK-7	.023-.028″
K-9	.023-.028″
UK-10	.023-.028″
K-54R	.013-.018″
K-57R	.018-.023″
K-59T	.015-.019″
K-60R	.018-.023″
K-79T	.013-.017″
K-80R	.013-.017″
K-83R	.013-.017″

L	
L-4J	.028-.033″
L-5	.018-.023″
L-7	.023-.028″
UL-12Y	.028-.033″
L-54R	.011-.014″
L-57R	.018-.023″
L-60R	.018-.023″
L-61Y	.018-.023″
L-62R	.018-.023″
L-64Y	.018-.023″
L-66Y	.023-.028″
L-81	.023-.028″
L-82Y	.023-.028″
L-83R	.011-.014″
L-84R	.011-.014″
L-85	.023-.028″
L-86	.018-.023″
L-87R	.011-.014″
L-87Y	.028-.033″

N	
N-3	.018-.023″
N-4	.023-.028″
N-5	.023-.028″
N-6Y	.023-.028″
N-9Y	.023-.028″
N-52R	.013-.018″
N-54R	.018-.023″
N-57	.018-.023″
N-57R	.018-.023″
N-60R	.018-.023″
N-60Y	.028-.033″
N-62R	.018-.023″
N-63Y	.028-.033″
N-64Y	.028-.033″
N-65Y	.028-.033″
N-66Y	.028-.033″
N-81R	.015-.018″
N-83R	.015-.018″
N-77R	.013-.018″

R	
R-52R	.015-.018″
R-54R	.015-.018″
R-56R	.015-.018″
R-59R	.015-.018″
R-61	.015-.018″
R-63	.015-.018″

appendix B

Combustion Chamber Modification Diagrams

This section includes several drawings which have been used earlier in the book in a reduced form. We have shown them actual size here, so that they may be traced and used as templates in the workshop, when carrying out modifications to the head.

Reproduced from Page 108

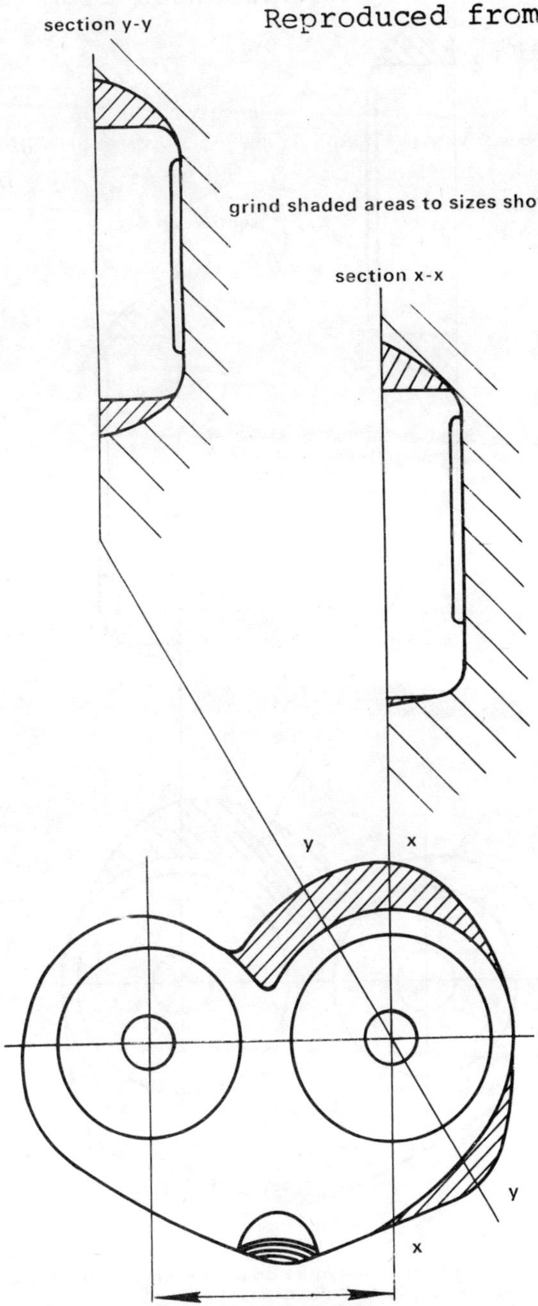

Fig 59 850 Mini chamber shape for use with standard 850 size valves only

Reproduced from Page 111

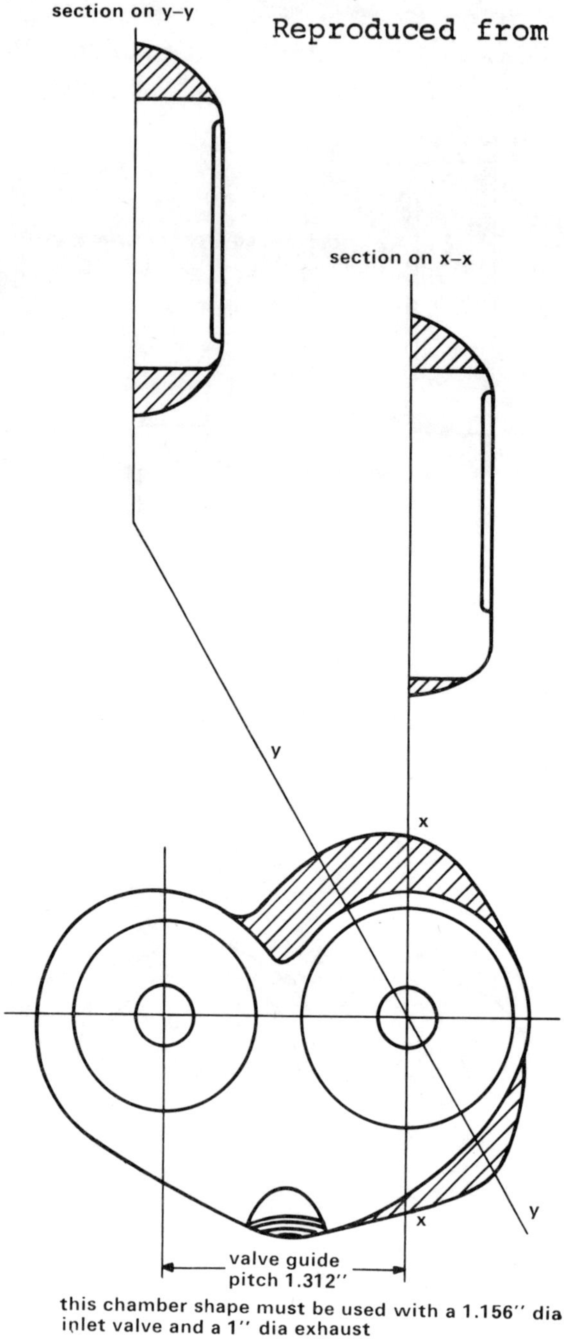

Fig 62 Chamber shape for use on 850, 948, 997 and 1100 cc engines
All other details as for figure 59

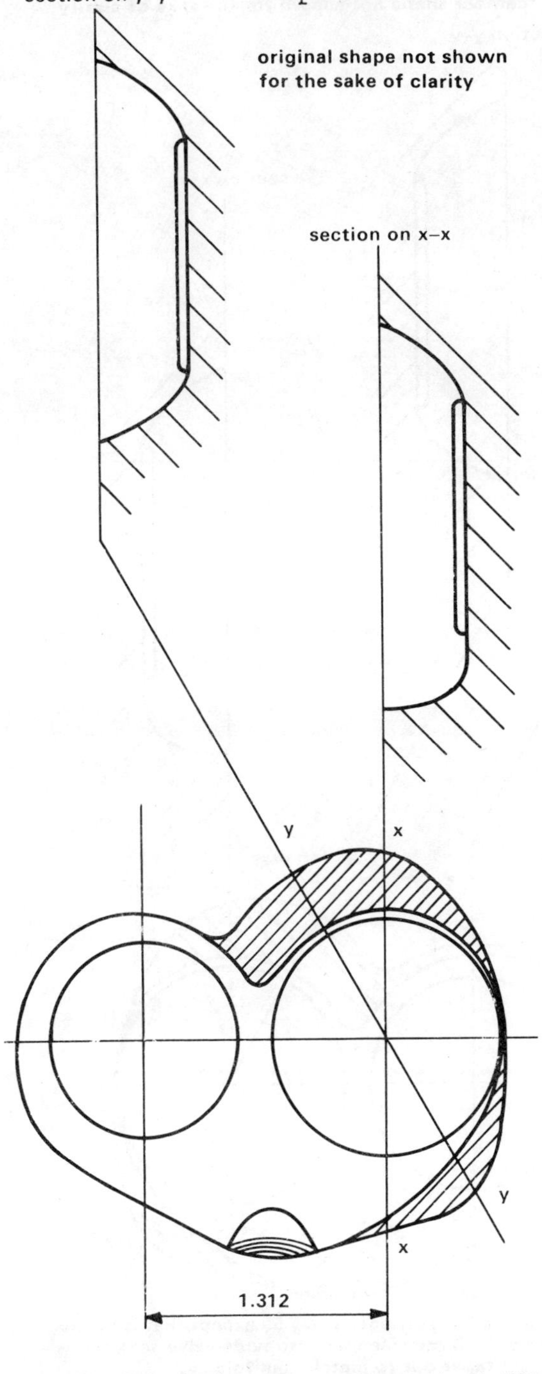

Fig 63 This chamber shape must be used with 1¼″ inlet and 1″ exhaust valves

Reproduced from Page 116

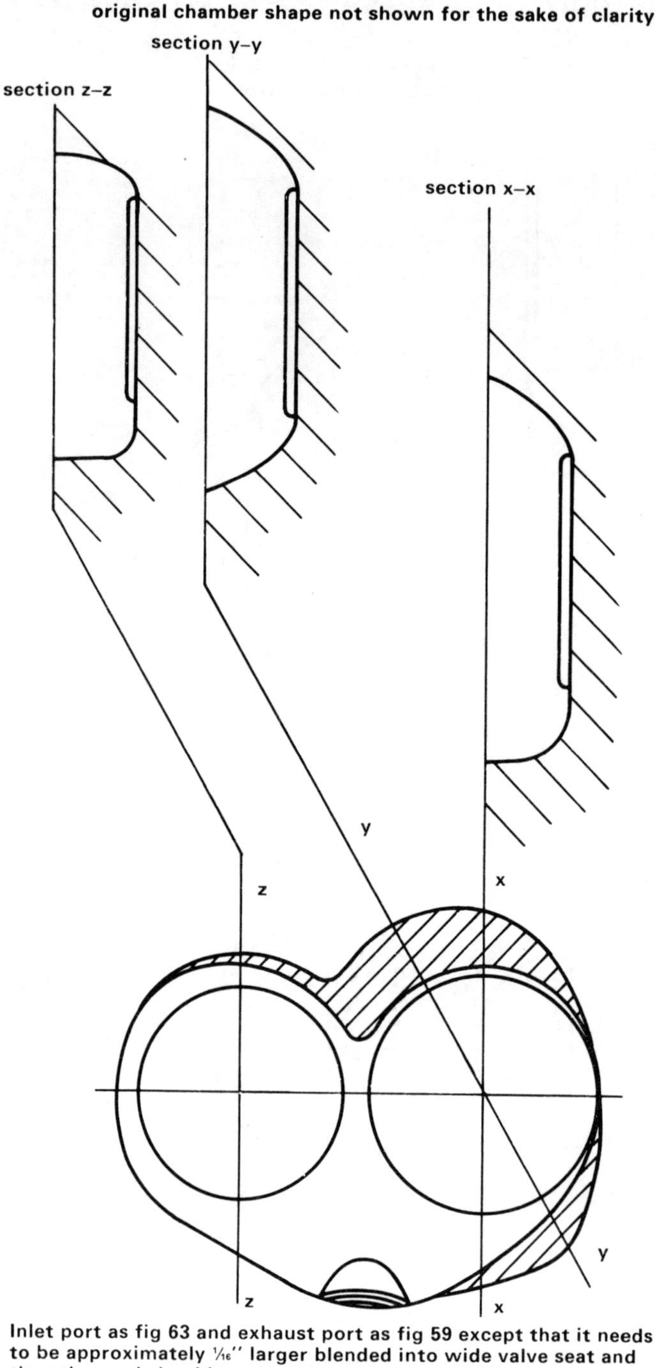

Inlet port as fig 63 and exhaust port as fig 59 except that it needs to be approximately 1/16" larger blended into wide valve seat and the other end should taper out to match manifold

Fig 65 This chamber shape must be used with 1¼" inlet valves and 1 1/32" exhaust valves

Reproduced from Page 118

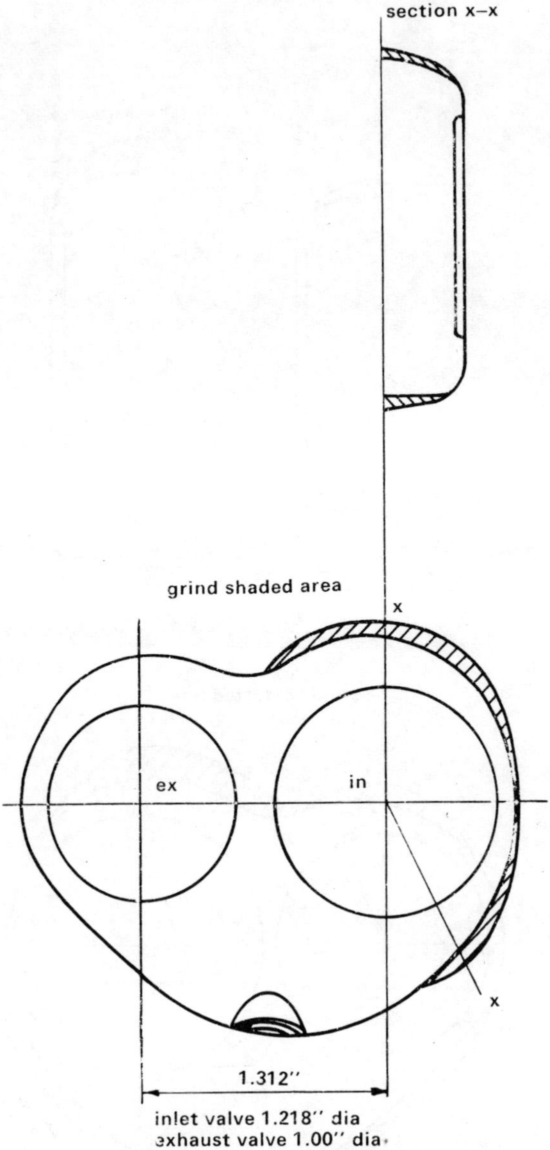

Fig 67 Chamber shape for 998 Cooper, 1100 MG and Sprites and Midgets from 1964 onwards

Reproduced from Page 120

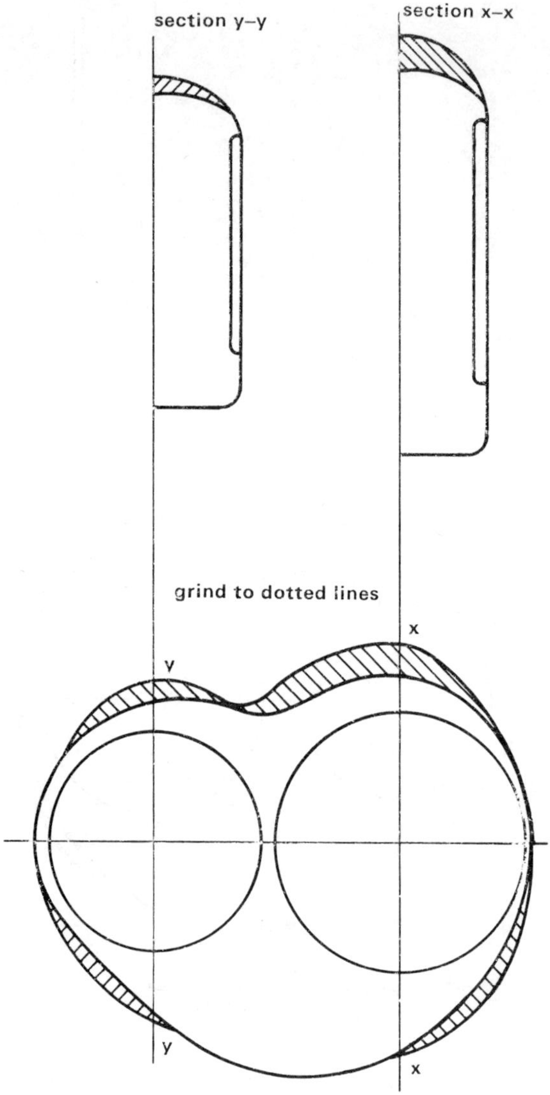

Fig 68 Full race head for 850, 948, 998 and 1098 cc

Reproduced from Page 122

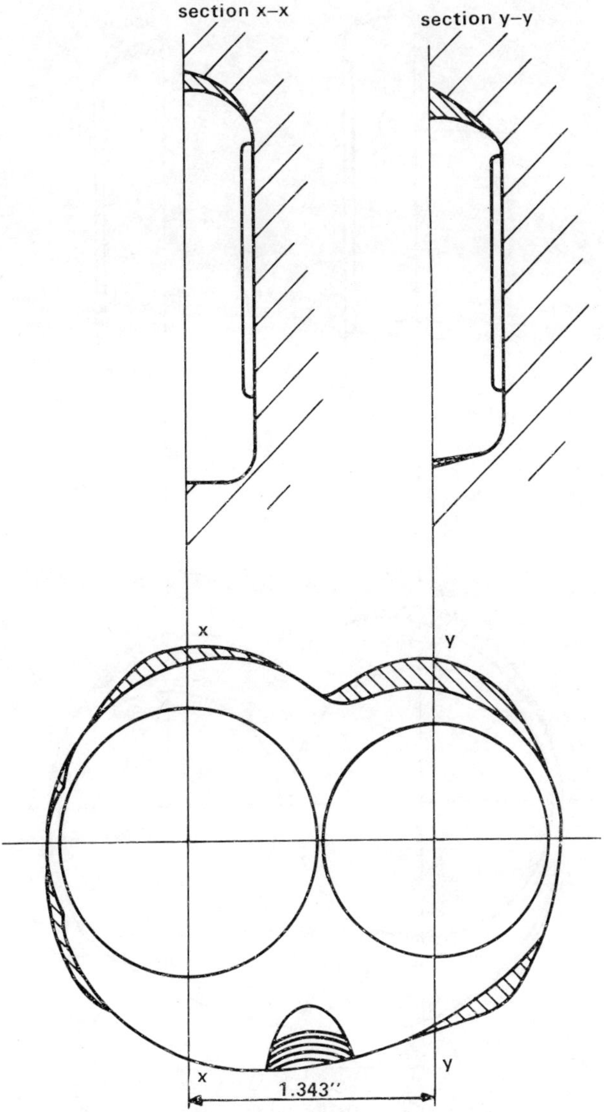

reshape inlet port as in fig 68 smooth and taper out exhaust port to match gasket and manifold

Fig 69 Chamber shape for Cooper S heads

Reproduced from Page 123

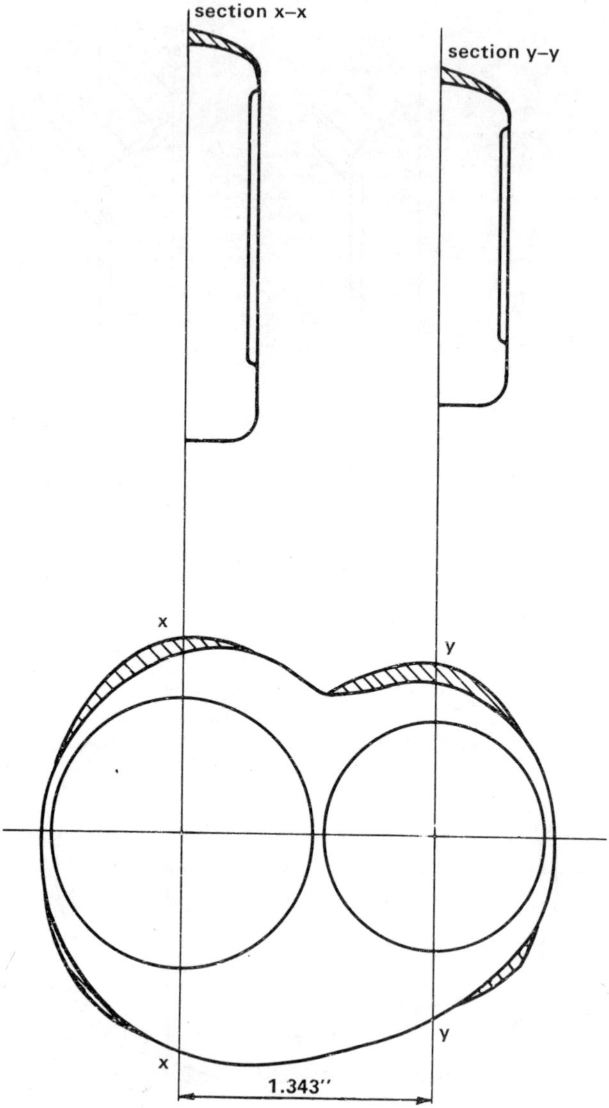

inlet port as in fig 68 smooth and taper out exhaust port to match gasket and manifold

Fig 70 Chamber shape for 1275 Sprite/Midget/1300 heads using 1.406" inlet and 1.156" exhaust

Reproduced from Page 124

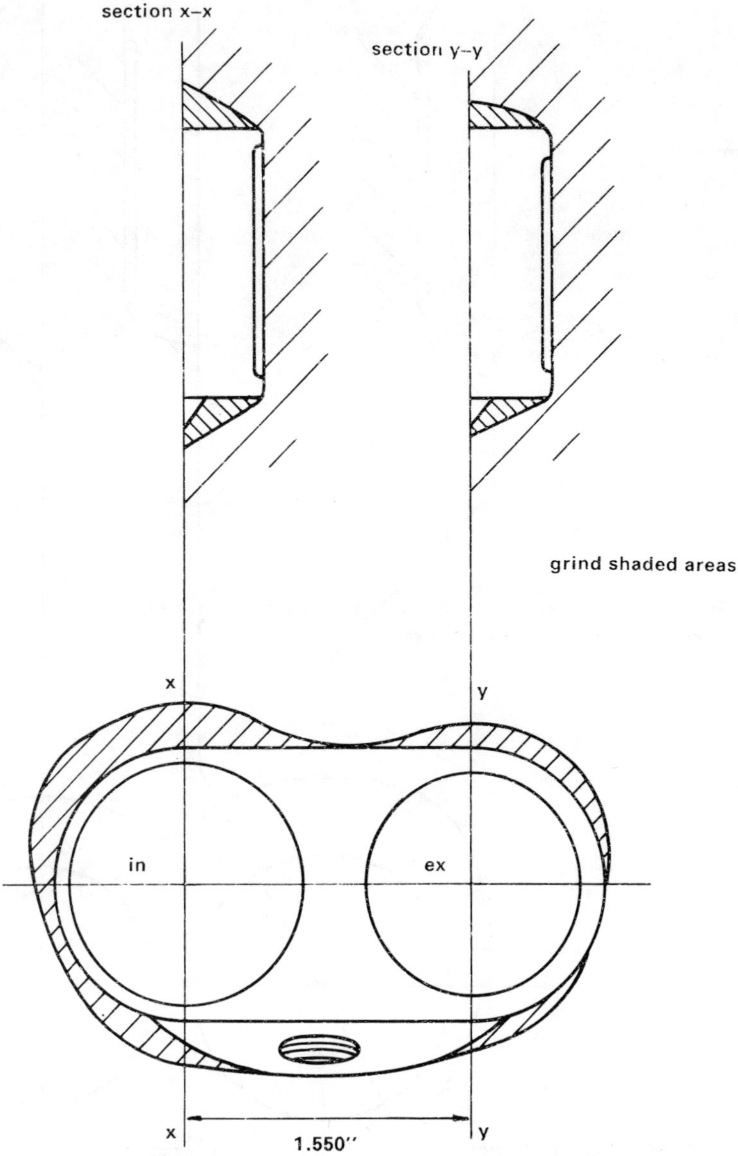

Fig 71 Standard valved 105E combustion chamber

Reproduced from Page 126

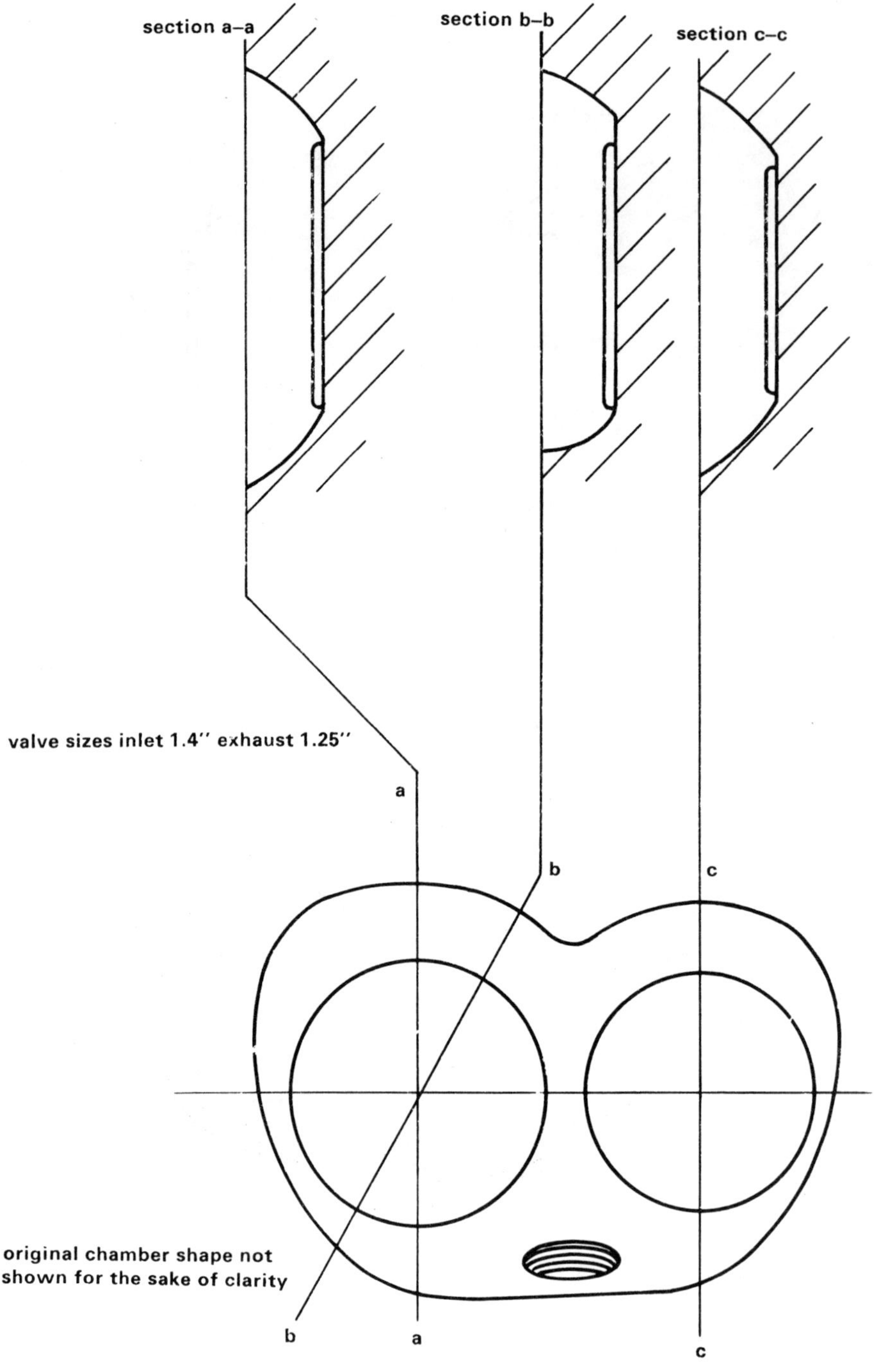

Fig 72 Chamber shape for big valve 105E head

Reproduced from Page 129

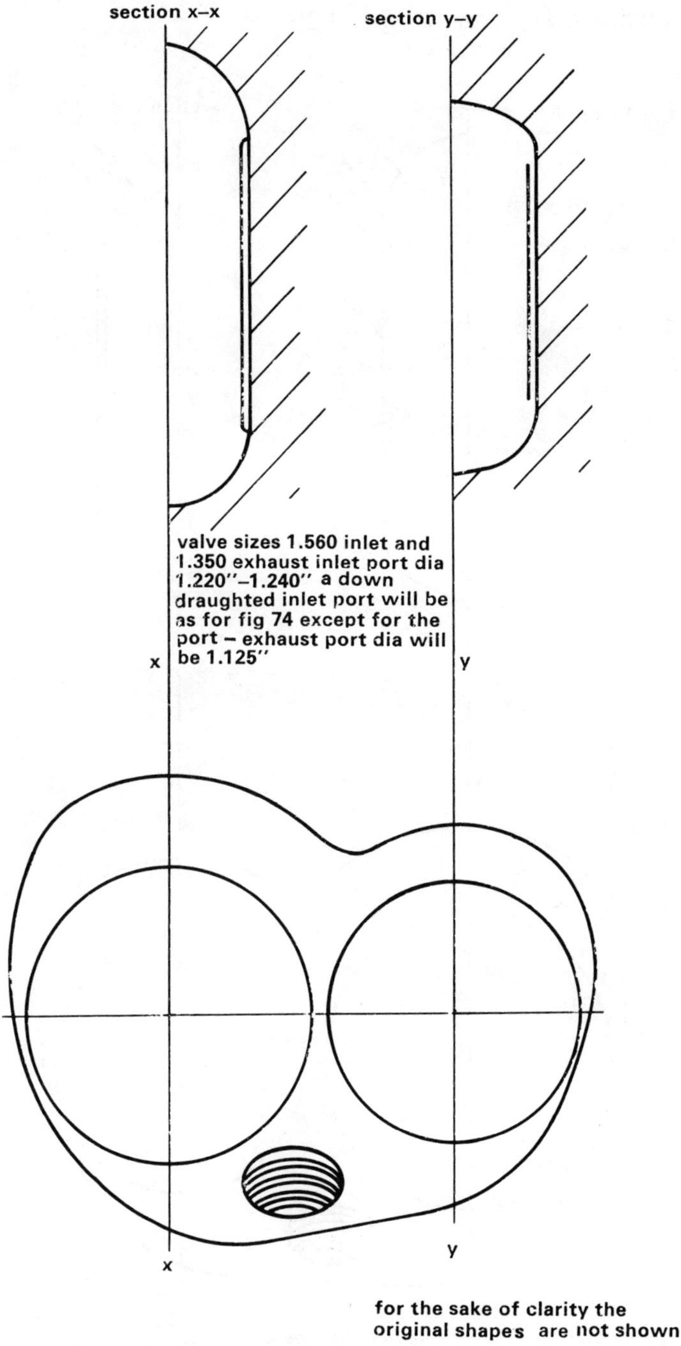

Fig 75 Chamber shape A for 1500 GT head

Reproduced from Page 129

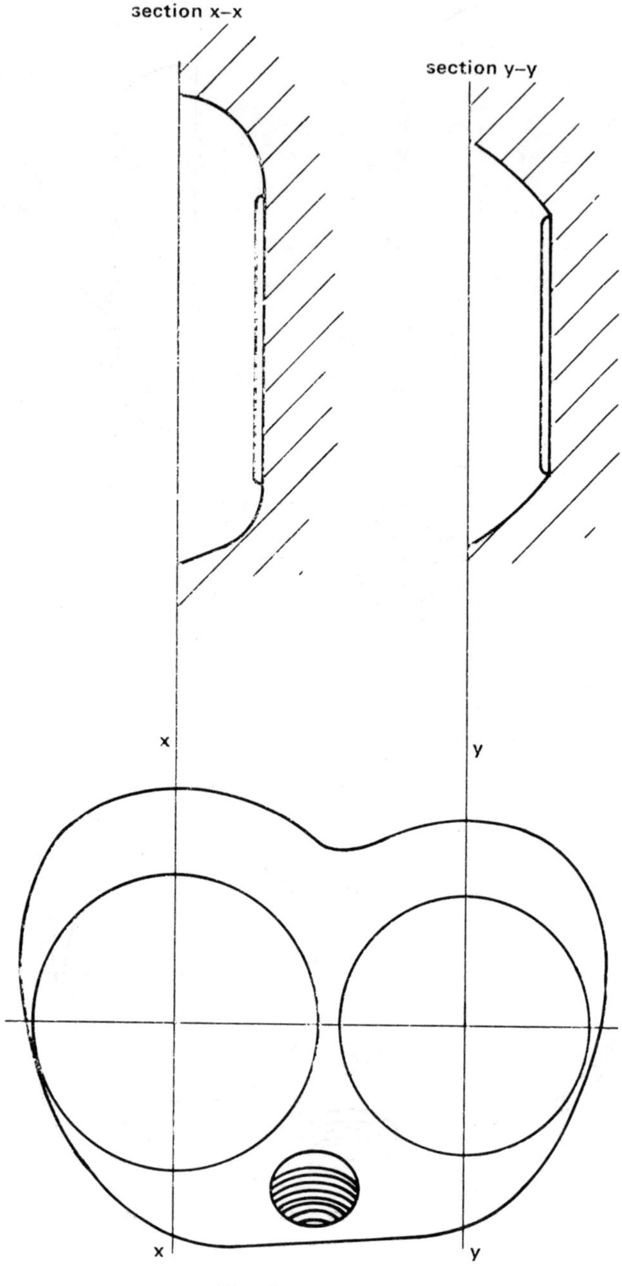

Chamber shape B

Reproduced from Page 131

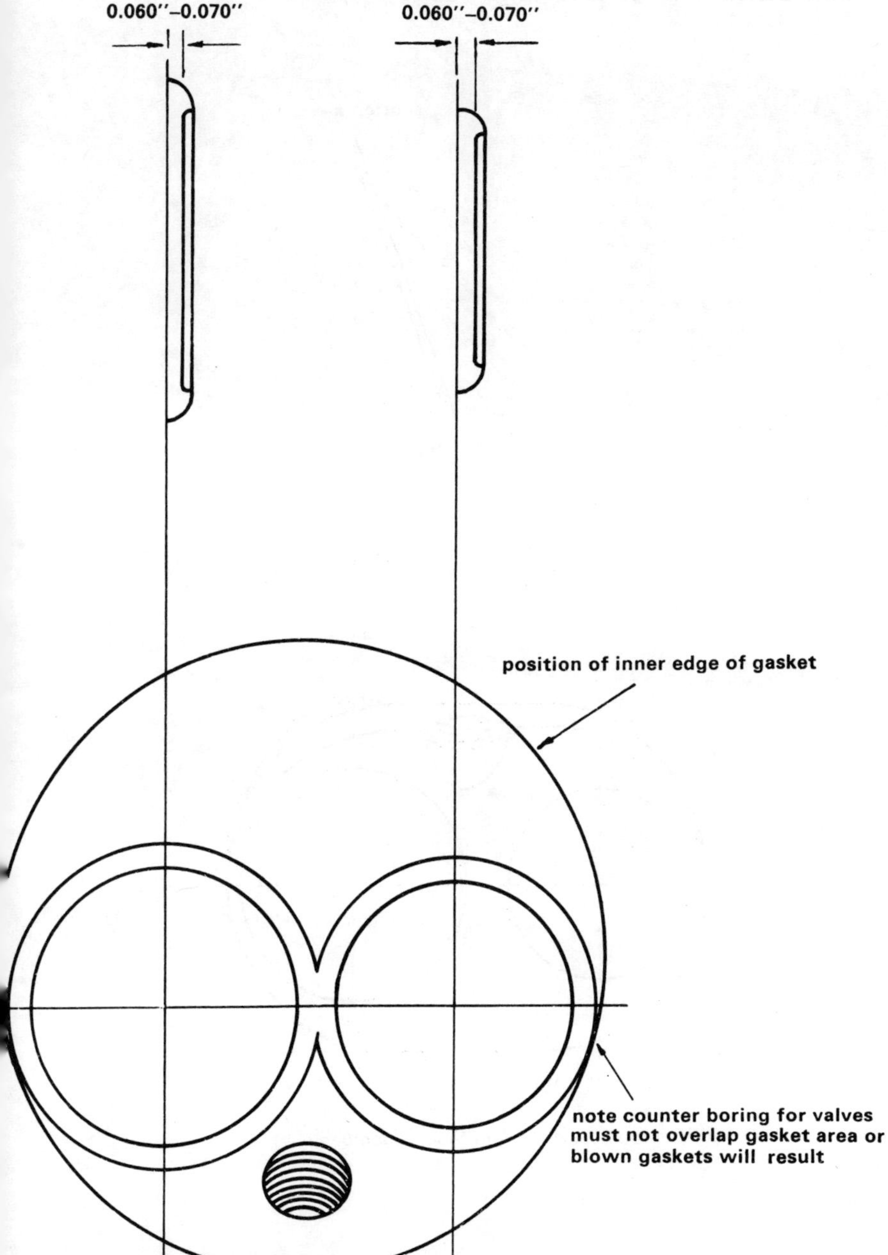

76 Combustion chamber for cross-flow Ford 1100 and 1300 non GT head modified to take GT valves

Reproduced from Page 133

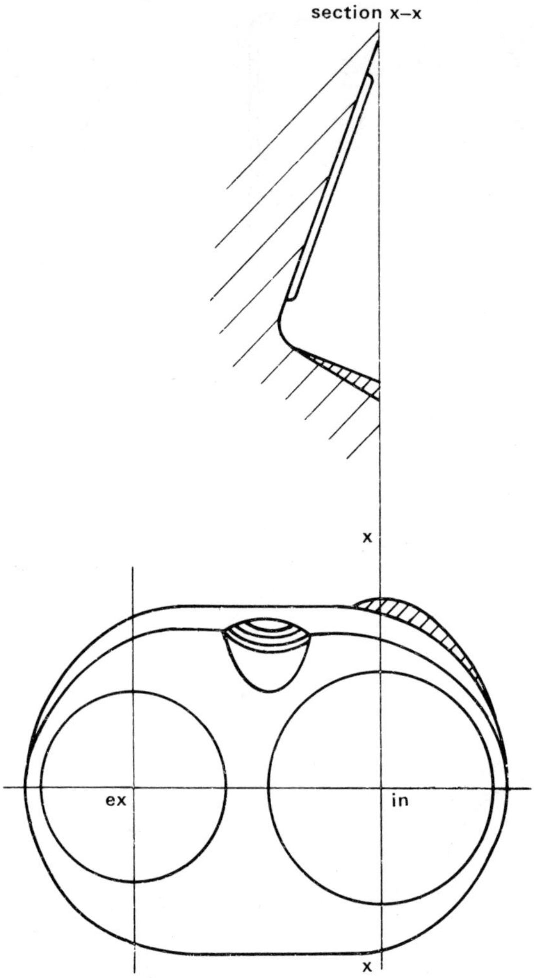

Fig 77 Modification to Imp Sport Chamber

Reproduced from Page 135

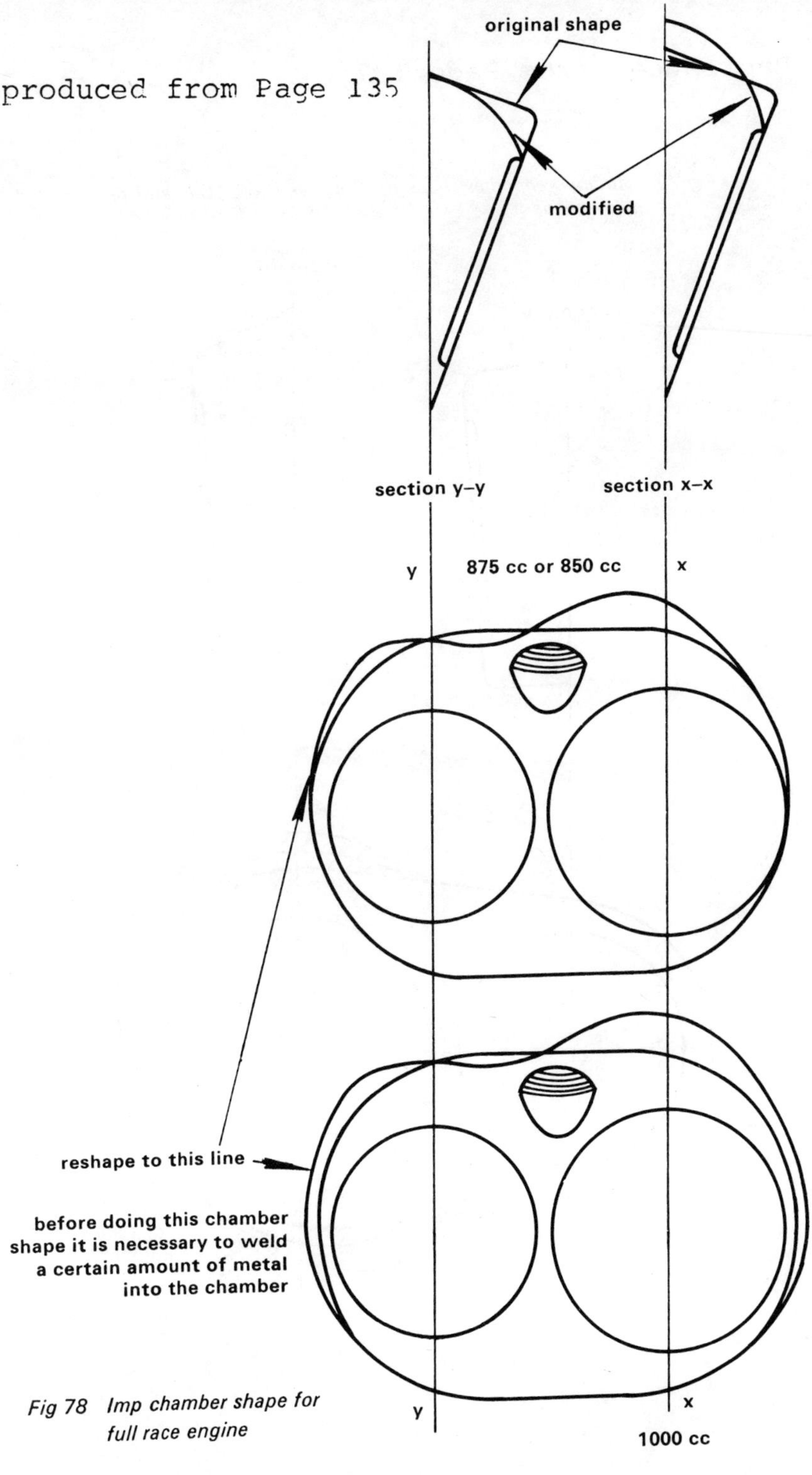

Fig 78 Imp chamber shape for full race engine

Reproduced from Page 137

Fig 79 Modified chamber shape for Chrysler 1500, 1600 and 1725 alloy heads

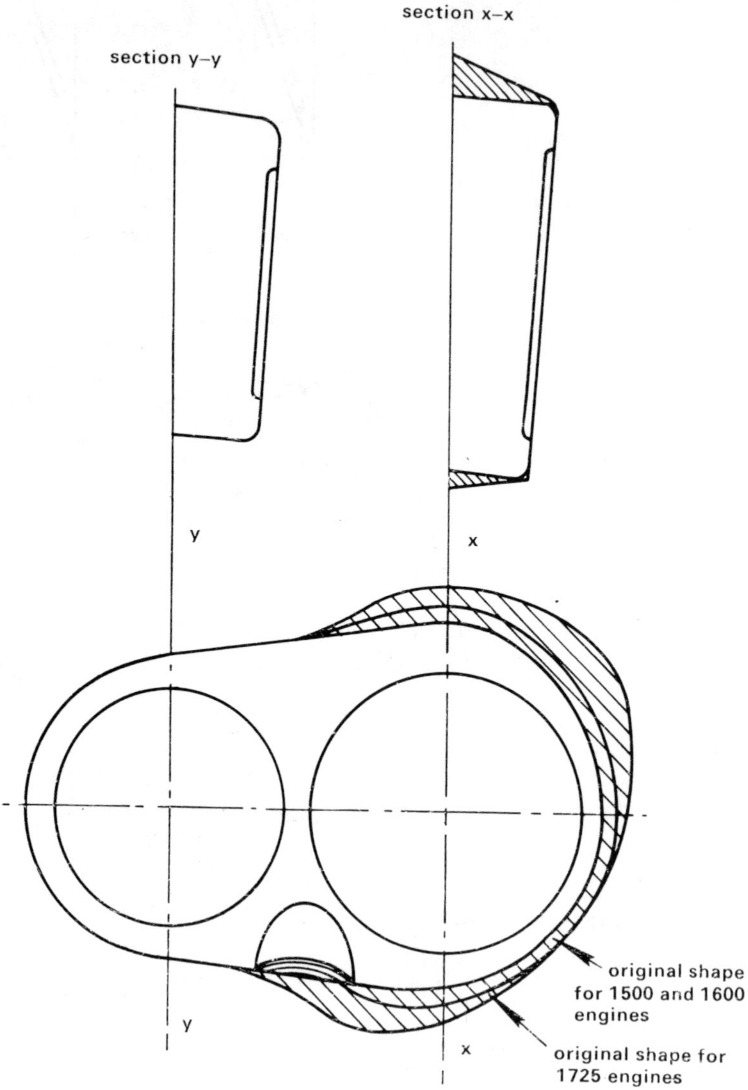

Reproduced from Page 137

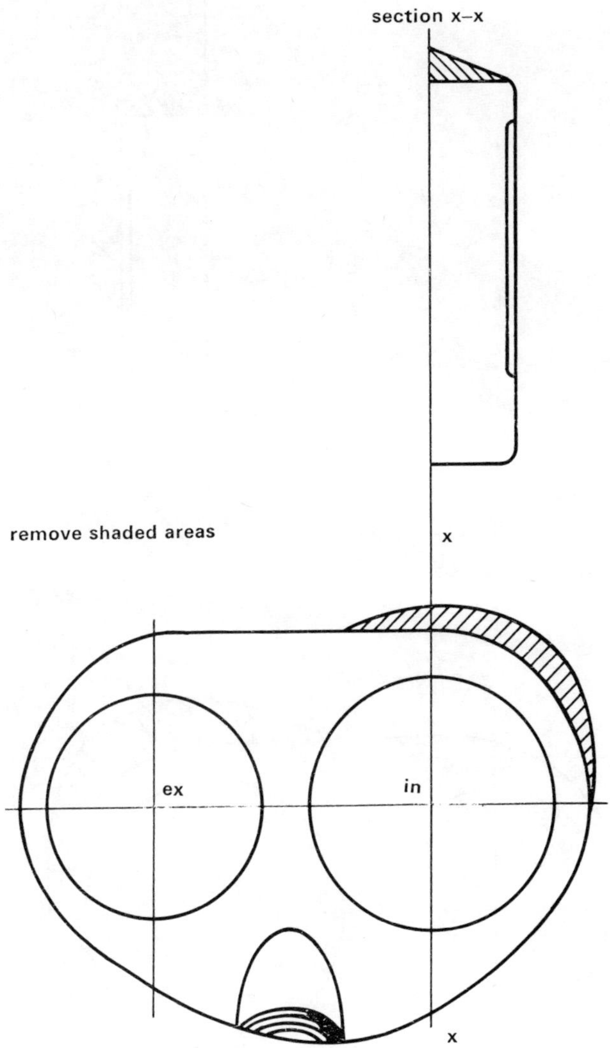

inlet and exhaust ports should have all irregularities blended out but must not be enlarged

Fig 80 Chamber shape for 1390 Minx head

Reproduced from Page 138

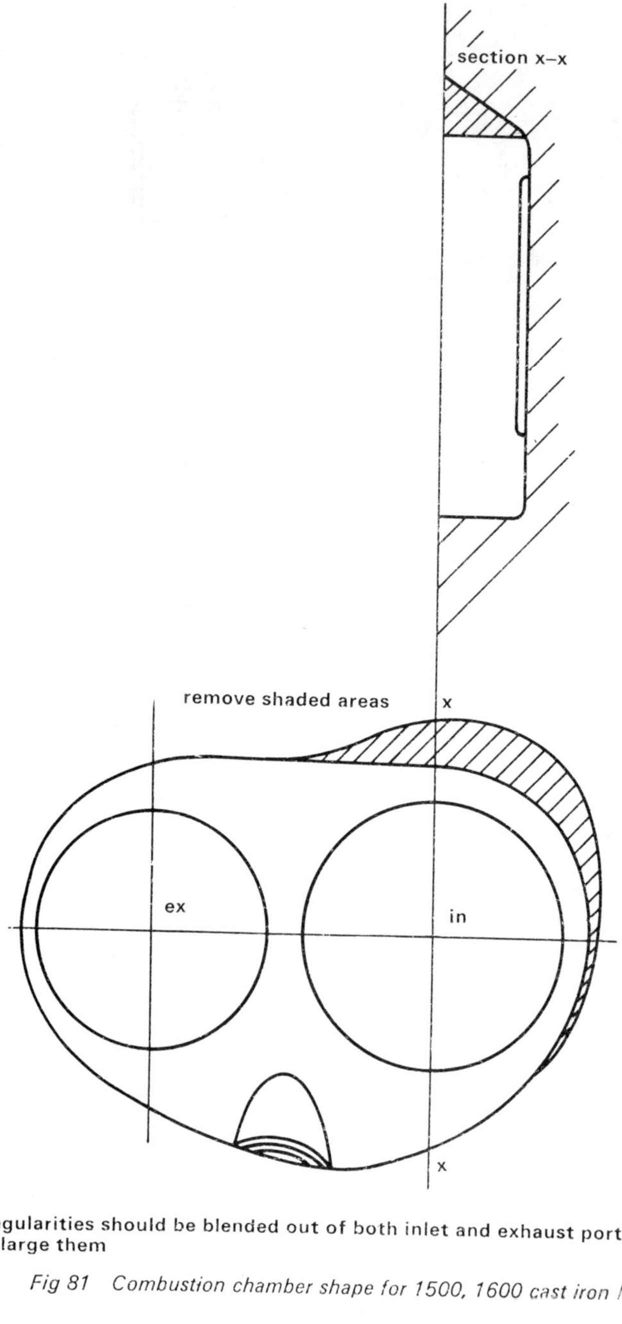

all irregularities should be blended out of both inlet and exhaust ports but do not enlarge them

Fig 81 Combustion chamber shape for 1500, 1600 cast iron Minx etc

Reproduced from Page 138

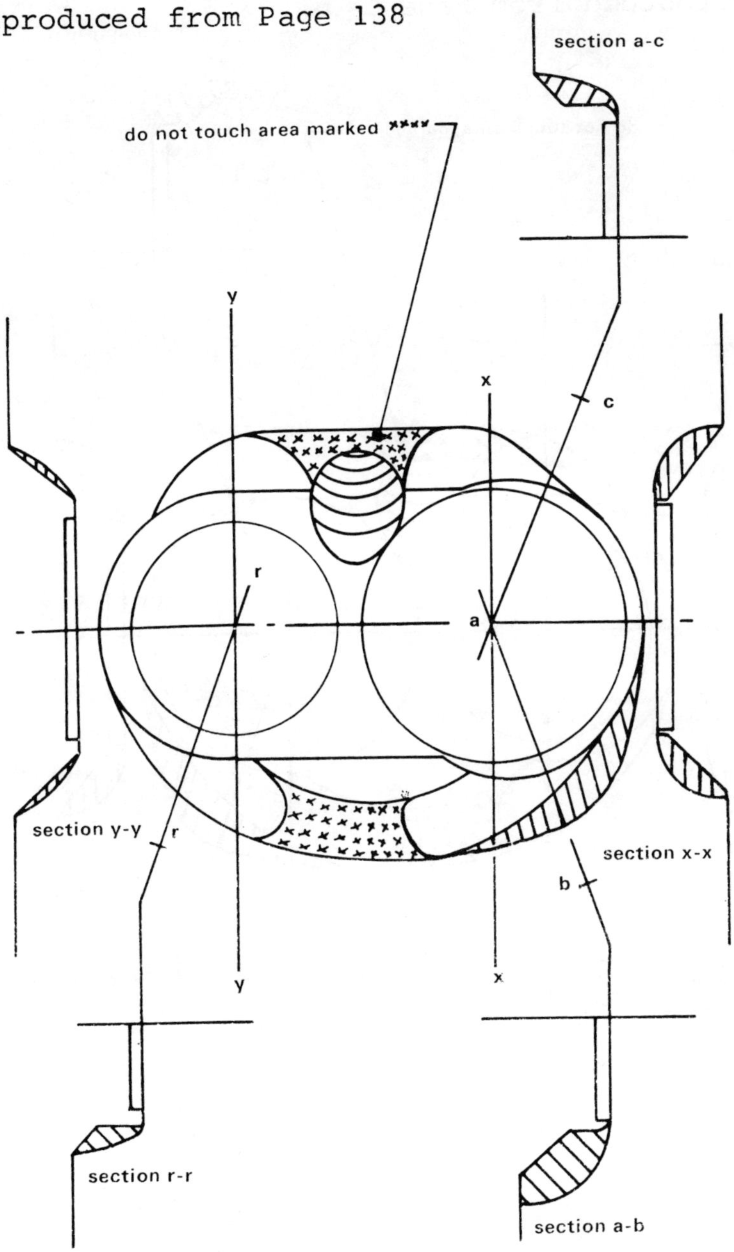

Fig 81 Combustion chamber shape for 1500, 1600 cast iron Minx etc

Reproduced from Page 140

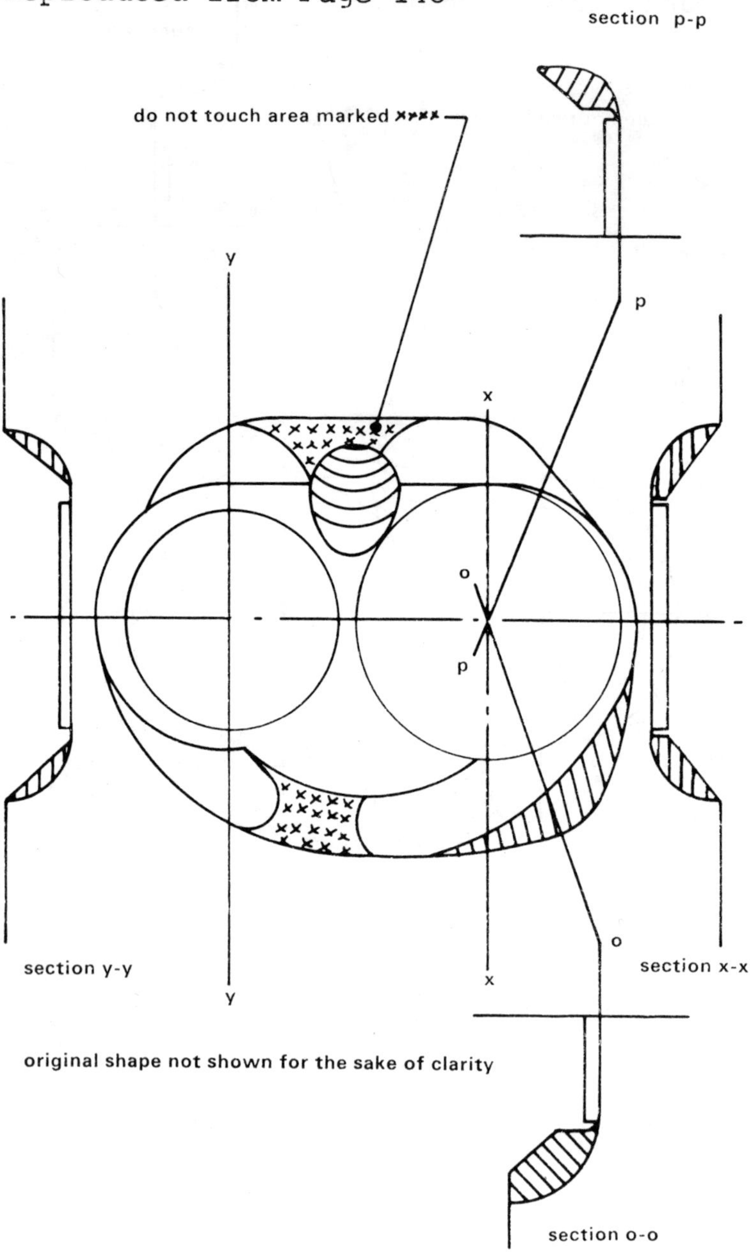

Fig 83 Port shape shown is finished profile, minimum enlargement should be used to achieve this shape

Reproduced from Page 142

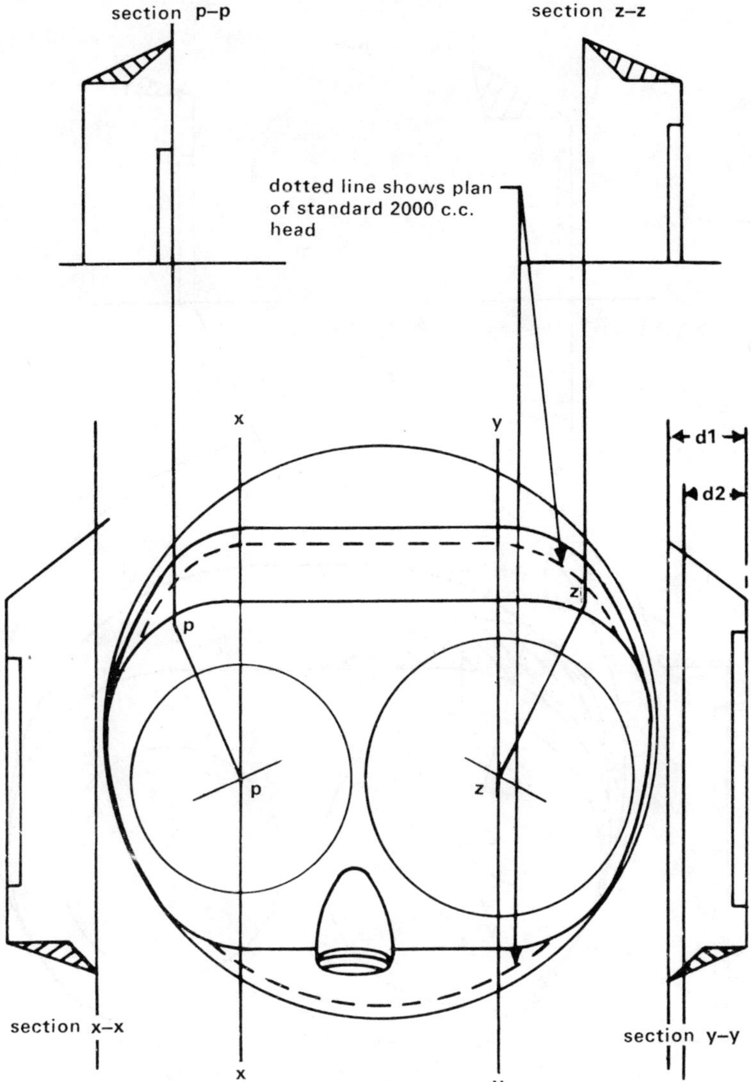

remove shaded areas

d1—standard chamber depth for 2.5 litre engines

d2—standard chamber depth for 2 litre engines

Fig 84 *Chamber modifications for TR5 and 6, 2500 P1, Mk 2 GT6 later type Triumph 2000 and Mk 2 2-litre Vitesses*

Reproduced from Page 144

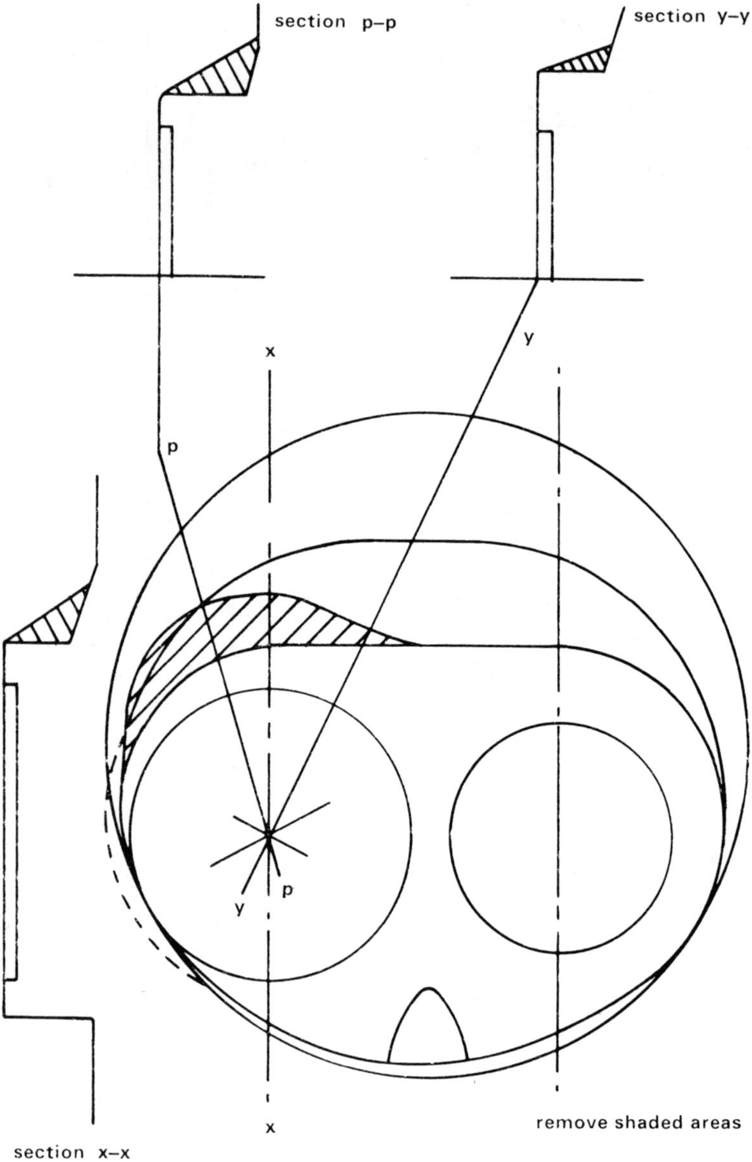

Fig 85 Chamber port modifications to TR4 and TR4A

LATEST PRICES - effective 1.8.1984

Catalogue No.		
020702	Tuning Volkswagen	£4.95
050036	Tuning Engines and Transmissions	£2.50
040050	Auto Electrics	£3.95
040089	Tuning Solex Carburetors	£3.95
020270	Tuning Anglias and Cortinas	£2.95
040086	Weber Carburetors Tuning and Maintenance	£3.95
040049	Tuning Lucas Ignition Systems	£4.95
020472	Tuning Four Cylinder Fords	£4.95
040091	Weber Carburetors Theory	£3.95
050018	Modifying Production Cylinder Heads	£2.95
050033	Theory and Practice of Chassis Tuning	£3.95
020508	More Mini Tuning	£3.50
040090	Tuning Stromberg Carburetors	£1.95
050034	Theory and Practice of Cylinder Head Modification	£5.95
040087	Tuning S.U. Carburetors	£4.95
050017	Introduction to Tuning	£2.50
020507	Tuning the Mini	£2.95
021466	Tuning Twin Cam Fords	£3.95
021465	Tuning Vivas and Firenzas	£3.95
020184	Tuning Datsuns	£4.95
040004	Car Test and Service Data	£5.95
040075	British Sölex carburetors	£5.95
040041	Bosch Electrical Systems	£4.95
040069	French Solex Carburetor Factory Manual	£4.95
040093	Zenith Carburetor Factory Manual	£4.95
170014	Motorcycling Engineering	£5.95
021385	Tuning Escort and Capris	£3.95
050114	Theory of Tuning	£3.95
040065	Engine and Electrical Service	£9.95